PROCEEDINGS OF SYMPOSIA IN APPLIED MATHEMATICS

VOLUME 1 NON-LINEAR PROBLEMS IN MECHANICS OF CONTINUA
Edited by E. Reissner (Brown University, August 1947)

VOLUME 2 ELECTROMAGNETIC THEORY
Edited by A. H. Taub (Massachusetts Institute of Technology, July 1948)

VOLUME 3 ELASTICITY
Edited by R. V. Churchill (University of Michigan, June 1949)

VOLUME 4 FLUID DYNAMICS
Edited by M. H. Martin (University of Maryland, June 1951)

VOLUME 5 WAVE MOTION AND VIBRATION THEORY
Edited by A. E. Heins (Carnegie Institute of Technology, June 1952)

VOLUME 6 NUMERICAL ANALYSIS
Edited by J. H. Curtiss (Santa Monica City College, August 1953)

VOLUME 7 APPLIED PROBABILITY
Edited by L. A. MacColl (Polytechnic Institute of Brooklyn, April 1955)

VOLUME 8 CALCULUS OF VARIATIONS AND ITS APPLICATIONS
Edited by L. M. Graves (University of Chicago, April 1956)

VOLUME 9 ORBIT THEORY
Edited by G. Birkhoff and R. E. Langer (New York University, April 1957)

VOLUME 10 COMBINATORIAL ANALYSIS
Edited by R. Bellman and M. Hall, Jr. (Columbia University, April 1958)

VOLUME 11 NUCLEAR REACTOR THEORY
Edited by G. Birkhoff and E. P. Wigner (New York City, April 1959)

VOLUME 12 STRUCTURE OF LANGUAGE AND ITS MATHEMATICAL ASPECTS
Edited by R. Jakobson (New York City, April 1960)

VOLUME 13 HYDRODYNAMIC INSTABILITY
Edited by R. Bellman, G. Birkhoff, C. C. Lin (New York City, April 1960)

VOLUME 14 MATHEMATICAL PROBLEMS IN THE BIOLOGICAL SCIENCES
Edited by R. Bellman (New York City, April 1961)

VOLUME 15 EXPERIMENTAL ARITHMETIC, HIGH SPEED COMPUTING, AND MATHEMATICS
Edited by N. C. Metropolis, A. H. Taub, J. Todd, C. B. Tompkins (Atlantic City and Chicago, April 1962)

VOLUME 16 STOCHASTIC PROCESSES IN MATHEMATICAL PHYSICS AND ENGINEERING
Edited by R. Bellman (New York City, April 1963)

VOLUME 17 **APPLICATIONS OF NONLINEAR PARTIAL DIFFERENTIAL EQUATIONS IN MATHEMATICAL PHYSICS**
Edited by R. Finn (New York City, April 1964)

VOLUME 18 **MAGNETO-FLUID AND PLASMA DYNAMICS**
Edited by H. Grad (New York City, April 1965)

VOLUME 19 **MATHEMATICAL ASPECTS OF COMPUTER SCIENCE**
Edited by J. T. Schwartz (New York City, April 1966)

VOLUME 20 **THE INFLUENCE OF COMPUTING ON MATHEMATICAL RESEARCH AND EDUCATION**
Edited by J. P. LaSalle (University of Montana, August 1973)

AMS SHORT COURSE LECTURE NOTES
Introductory Survey Lectures

VOLUME 21 **MATHEMATICAL ASPECTS OF PRODUCTION AND DISTRIBUTION OF ENERGY**
Edited by P. D. Lax (San Antonio, Texas, January 1976)

VOLUME 22 **NUMERICAL ANALYSIS**
Edited by G. H. Golub and J. Oliger (Atlanta, Georgia, January 1978)

VOLUME 23 **MODERN STATISTICS: METHODS AND APPLICATIONS**
Edited by R. V. Hogg (San Antonio, Texas, January 1980)

VOLUME 24 **GAME THEORY AND ITS APPLICATIONS**
Edited by W. F. Lucas (Biloxi, Mississippi, January 1979)

VOLUME 25 **OPERATIONS RESEARCH: MATHEMATICS AND MODELS**
Edited by S. I. Gass (Duluth, Minnesota, August 1979)

VOLUME 26 **THE MATHEMATICS OF NETWORKS**
Edited by S. A. Burr (Pittsburgh, Pennsylvania, August 1981)

VOLUME 27 **COMPUTED TOMOGRAPHY**
Edited by L. A. Shepp (Cincinnati, Ohio, January 1982)

VOLUME 28 **STATISTICAL DATA ANALYSIS**
Edited by R. Gnanadesikan (Toronto, Ontario, August 1982)

VOLUME 29 **APPLIED CRYPTOLOGY, CRYPTOGRAPHIC PROTOCOLS, AND COMPUTER SECURITY MODELS**
By R. A. DeMillo, G. I. Davida, D. P. Dobkin, M. A. Harrison, and R. J. Lipton (San Francisco, California, January 1981)

AMS SHORT COURSE LECTURE NOTES
Introductory Survey Lectures
published as a subseries of
Proceedings of Symposia in Applied Mathematics

PROCEEDINGS OF SYMPOSIA
IN APPLIED MATHEMATICS
Volume 30

POPULATION BIOLOGY

AMERICAN MATHEMATICAL SOCIETY
PROVIDENCE, RHODE ISLAND

LECTURE NOTES PREPARED FOR THE
AMERICAN MATHEMATICAL SOCIETY SHORT COURSE

POPULATION BIOLOGY

HELD IN ALBANY, NEW YORK
AUGUST 6–7, 1983

EDITED BY
SIMON A. LEVIN

The AMS Short Course Series is sponsored by the Society's Committee on Employment and Education Policy (CEEP). The series is under the direction of the Short Course Advisory Subcommittee of CEEP.

Library of Congress Cataloging in Publication Data

Main entry under title:

Population biology.

(Proceedings of symposia in applied mathematics, ISSN 0160-7634; v. 30. AMS short course lecture notes)

"Lecture notes prepared for the American Mathematical Society short course, population biology, held in Albany, NY, August 6–7, 1983"–T.p. verso.

Includes bibliographies.

Contents: Mathematical population biology/Simon Levin–Population dynamics and demography/James Frauenthal–Some mathematical problems in population genetics/Thomas Nagylaki–[etc.]

1. Population biology–Mathematical models–Congresses. I. Levin, Simon A. II. American Mathematical Society. III. Series: Proceedings of symposia in applied mathematics; v. 30. IV. Series: Proceedings of symposia in applied mathematics. AMS short course lecture notes.

QH352.P57 1984 575.5'248'0151 83-21389
ISBN 0-8218-0083-3

1980 Mathematics Subject Classification. Primary 92A15, 92A10, 92A17.

Printed in the United States of America.

This volume was printed directly from copy prepared by the authors.

CONTENTS

PREFACE

The lecture notes contained in this volume were presented at the AMS short course on population biology, held August 6-7, 1983 in Albany, New York in conjunction with the eighty-seventh summer meeting of the American Mathematical Society.

Population biology is probably the oldest area in mathematical biology, but remains a constant source of new mathematical problems and the area of biology best integrated with mathematical theory. The need for mathematical approaches has never been greater, as evolutionary theory is challenged by new interpretations of the paleontological record and new discoveries at the molecular level, as world resources for feeding populations become limiting, as the problems of pollution increase, and as both animal and plant epidemiological problems receive closer scrutiny.

The purpose of this course was to acquaint the participant with the mathematical ideas that pervade almost every level of thinking in population biology and to provide an introduction to the many applications of mathematics in the field.

Proceedings of Symposia in Applied Mathematics
Volume 30, 1984

MATHEMATICAL POPULATION BIOLOGY

Simon A. Levin[1]

INTRODUCTION

Mathematical population biology encompasses the modeling of a wide range of biological problems, including the growth of natural populations, changes in their demography and distribution, and alteration of their genotypic and phenotypic compositions.

There are a number of important historical lineages leading to the present state of the subject. Especially familiar to mathematicians are the works of Volterra, Lotka, and Kostitzin (see Scudo and Ziegler, 1978) dealing with the dynamics of populations, and the fundamental writings of Fisher, Wright, and Haldane (see Provine, 1971) in theoretical population genetics. The contributions of these pioneers began with Volterra's 1900 speech at the University of Rome on attempts to apply mathematics in the biological and social sciences (Volterra 1901-2; see Borsellino 1980), and they have continued even until the present (see for example Wright, 1980, 1982). Work in epidemiology can also be traced to the same period (e.g., Brownlee 1906), and mathematical studies in genetics actually were initiated even earlier (e.g., Pearson 1896).

Of course, one can identify the impetus for quantitative demography as deriving from much earlier work, especially that of Malthus (1798), and even Malthus' writings were preceded by more than a century by the work of John Graunt (1662). Graunt studied the birth and death rates of the human population in London, from which he estimated quite closely the population growth rate. He calculated that, neglecting immigration, the population would double every 64 years. He further recognized that such a situation could not have persisted since the Creation and that therefore the simplest exponential growth model was inadequate. However, it was only somewhat later, and inspired by Malthus, that such observations found expression in density-dependent population models. To account for density effects, Sadler (1830) proposed a

1980 Mathematics Subject Classification. 92A10, 92A15, 92A17
[1]Supported by National Science Foundation Grant MCS-8001618

model in which fecundity was inversely related to density; another approach was due to Quetelet (1835) and Fourier (perhaps slightly earlier than Quetelet) proposed purely phenomenological models, analogized from experience with hydrodynamic forces. Quetelet introduced the notion of "resistance to growth," which was proportional to the square of the rate of increase; but this approach was not related to any specific biological mechanism (Hutchinson, 1978).

Finally (Hutchinson, 1978), Pierre Francois Verhulst (1838, 1845), motivated by his colleague Quetelet, proposed the now famous logistic equation in which the per capita rate of increase of population growth is assumed to be a linearly decreasing function of population size. Numerous other density-dependent population models have been put forth since, but the logistic is the simplest which captures the features of a threshold population size, the "carrying capacity," above which population growth is negative due to intrapopulation inference and below which growth is positive. Frauenthal, in his contribution to the present volume, discusses in further detail the logistic and other density-dependent models. For a fuller discussion of the discrete time analogues of these population growth models, one is referred to May (1978) and Levin and Goodyear (1980). Such discrete time models are especially appropriate to populations which do not breed continuously and are widely used in fisheries science (Ricker, 1958).

USES OF MODELS

It is important to understand the manifold purposes to which mathematical models are put, and to realize that models constructed for one purpose can neither be indiscriminately applied to other uses, nor judged by criteria other than those appropriate to the intended use. Despite the need for predictive capacity in evaluating the potential effects of human influences on natural populations, most population models cannot meet the criteria necessary for long-term prediction. Their general inadequacy arises because of the cumulative influence of density-independent factors such as climatic fluctuations, which introduce unpredictable parametric variation into models; the propensity for many nonlinear models to propagate errors and to exhibit sensitivity to certain parameters; and the inherent difficulties in estimating parameters from data.

However, there are some models available, for example the forest growth simulators of Botkin et al. (1972), which have been shown to be quite reliable, at least for limited time horizons. In general, those models that perform best recognize the stochastic nature of population processes, generating probability distributions rather than deterministic predictions, and that bypass the critical stock-recruit relation, which plagues fishery science, by considering the number of new recruits added to the population each year to be independent of population size. This is quite appropriate in forest growth models, especially when the spatial extent is limited and/or when the time period is not so long as to exhaust the stored supply of available new recruits. Forests accumulate and store potential recruits by burying seeds or by maintaining young individuals in the understory, ready to take over when older trees die.

Fish populations do not have such storage mechanisms. On the other hand, some fish species do buffer themselves against some of the unpredictability of nature by iteroparity (multiple spawning) and by producing huge excesses of eggs. Thus, a case can be made in some situations, especially when the spawning stock has not been severely depleted, for ignoring the details of the stock-recruit relationship and treating population size as a random variable. However, the appropriateness of this assumption and the limits of its applicability represent central and hotly-debated issues in fisheries ecology.

Most mathematical models are not intended for prediction, and in fact may or may not represent hypotheses concerning how real populations are regulated. Models may be used to explore the consequences of particular, restrictive assumptions which represent only part of the full picture; or may be used to identify critical system components, linkages or parameters to be measured; or may be used as mechanisms by which one identifies critical experiments, or simply fixes ideas. (Levin 1980b, 1983a; Strong 1983).

For example, in fisheries biology, it is well-recognized that the simplest discrete-time population models have the potential for exhibiting, with respect to parametric variation, bifurcations to periodic solutions (Ricker 1954, May, 1974, 1978) or much more complicated behavior (Li and Yorke 1975, May 1974), and that there are serious problems regarding the problem of parameter identification. Thus, and because even the correct form of such models is not known, they are inadequate for predicting long-term population trends. However, they have been extremely useful as pedagogical tools and for developing theoretical constructs. For example, in epidemiology, a wide class of models show the existence of a threshold population size for the establishment of a disease (Anderson 1982). Different models may make different quantitative predictions concerning the numerical value of the threshold; but consideration of a variety of such models may give order of magnitude approximations or bounds, and even the demonstration that there exists a threshold represents a major theoretical contribution.

CONTENTS OF SUBSEQUENT PAPERS

This volume includes contributions by experts regarding a selection of topics of current interest in population biology. In such a short collection, it is impossible to be complete, and the reader is referred elsewhere for further depth. General references for the application of mathematics in population biology are May (1974, 1978), Levin (1978), and Roughgarden (1979); basic references in population genetics include Ewens (1979), Nagylaki (1977), and Crow and Kimura (1970).

The opening two contributions to the short course were Jim Frauenthal's discussion of demographic theory (for other discussions see Nisbet and Gurney (1982), Keyfitz (1977), Smith and Keyfitz (1977), and Impagliazzo (1984)), and James Yorke's lecture on models of gonorrhea. Unfortunately, lecture notes for Professor Yorke's lecture were not available for these proceedings, but the reader interested in a fuller discussion will find it in Hethcote and Yorke (1984). Basic treatises on epidemological models are Bailey (1957), Hoppensteadt (1975), Frauenthal (1980), and Anderson (1982). Demographic theory is at the heart of mathematical population biology, and consideration of the dynamics of infectious diseases represents an area of application in which theory and data have been brought into close contact.

The second pair of lectures, by Thomas Nagylaki and Ethan Akin, reflect several active aspects of modern evolutionary theory, ranging from explicit population genetic models (including very current work on gene conversion) to game-theoretic phenomenological approaches.

Evolutionary theory is concerned on the one hand with the patterns of change in phenotypic distributions (e.g., observed characters) and secondly with the genetic mechanisms which govern that change. The grand synthesis which took place in the 1940's served to meld these approaches; but what this synthesis actually meant is interpreted differently by different people and is in need of constant reassessment as our knowledge of the molecular processes changes (Mayr and Provine 1980).

Many ecologists use optimization and adaptation interchangeably and assume that these are the obvious endpoints of natural selection. But natural selection is an on-going, ever-changing processes in a changing environment (Lewontin 1977, Gould 1977, Levin 1978, 1980a), and the <u>process</u> of optimization does not guarantee an <u>optimal</u> end-product; nor is the notion of end-product really appropriate to such a "game with Nature" in which the only real payoff is being allowed to continue in the game (Slobodkin 1964). Furthermore, much phenotypic expression is not under direct genetic control, and this is another obstacle to be confronted by the adaptationist (Gould and Lewontin 1979).

The sense in which natural selection may be thought to optimize anything at all is at the core of evolutionary theory (see for example Levin 1978, Maynard Smith 1982) and is explored further in Ethan Akin's contribution to this volume. In particular, the famous theory (Fisher 1983) that the mean fitness of a population increases under natural selection, eventually reaching some "adaptive" peak, is a very special result, valid only for situations in which fitnesses are approximately constant. For most ecological situations, the notion of constant fitnesses is not appropriate, and a modified approach is needed (Levin 1978, Maynard Smith 1982).

The last two papers in the volume explore two other areas of active current interest. Wayne Getz discusses the application of control theory to the management of renewable resources (see also Clark 1976), and George Sugihara presents results on the application of graph theoretical methods to interpreting patterns in the structure of ecological food webs (see also Cohen 1978, Pimm 1982). Especially since Clark's 1976 book, bioeconomic problems have been recognized as a rich source of mathematical problems, and as constituting an area of great applied importance as human demands on renewable resources increase. The investigation of food web structure also represents an area which is quickly becoming one of the most active in ecological theory. It is appealing both for its intellectual content, and its potential applied importance as more and more attention is paid to possible ecosystem consequences of anthropogenic perturbations.

There are a number of other important topics not treated in this volume, for example host-parasite systems (Anderson 1982), coevolutionary models (Anderson and May 1983, Levin 1983b), and models of dispersal and spatial patterning (Skellam 1951, Okubo 1980, Levin 1974, 1983). It is hoped that those topics that are represented here will be sufficient to provide an entree to this exciting area, and that the additional references given will provide the opportunity for delving further into the subject.

REFERENCES

Anderson, R. M. (ed.). 1982. The Population Dynamics of Infectious Diseases: Theory and Applications. Chapman and Hall, London.

Anderson, R. M. and R. M. May. 1983. Epidemiology and genetics in the coevolution of parasites and hosts. To appear. Phil. Trans. Roy. Soc. B.

Bailey, N. T. J. 1957. The Mathematical Theory of Epidemics. Griffin, London.

Borsellino, A. 1980. Vito Volterra and contemporary mathematical biology. pp. 410-417. In C. Barigozzi (ed.) Vito Volterra Symposium on Mathematical Models in Biology. Springer-Verlag, Berlin, Heidelberg, New York.

Botkin, D. B., J. F. Janak, and J. R. Wallis. 1972. Some ecological consequences of a computer model of forest growth. J. Ecol. 60:849-872.

Brownlee, T. 1906. Statistical studies in immunity: the theory of an epidemic. Proc. Royal Society, Edinburgh. 26:484-521.

Clark, C. W. 1976. Mathematical Bioeconomics. The Optimal Management of Renewable Resources. Wiley, New York.

Cohen, J. E. 1978. Food Webs and Niche Space. Princeton University Press, Princeton.

Crow, J. K. and M. Kimura. 1970. An Introduction to Population Genetics Theory. Harper and Row, New York.

Ewens, W. 1979. Mathematical Population Genetics. Springer-Verlag, Berlin, Heidelberg, New York.

Fisher, R. A. 1930. Genetical Theory of Natural Selection. Oxford University Press, Oxford.

Frauenthal, J. C. 1980. Mathematical Modelling in Epidemiology. Springer-Verlag, Berlin, Heidelberg, New York.

Gould, S. J. and R. C. Lewontin. 1979. The spandrels of San Marco and the Panglossian paradigm: a critique of the adaptationist programme. Proc. R. Soc. B. 205:581-598.

Gould, S. J. 1977. Ever Since Darwin. Norton, New York.

Graunt, J. 1662. Natural and political observations mentioned in a following index, and made upon the Bills by Mortality of John Graunt Citizen of London. (reprinted in part in Smith and Keyfitz, Mathematical Demography).

Hethcote, H. and J. Yorke. 1984. Gonorrhea transmission dynamics and control. To appear. Lecture Notes in Biomathematics. Springer-Verlag, Berlin, Heidelberg, New York.

Hoppensteadt, F. 1975. Mathematical theories of populations: demographics, genetics and epidemics. Society for Industrial and Applied Mathematics. Philadelphia.

Hutchinson, G. E. 1978. An Introduction to Population Ecology. Yale University Press, New Haven.

Impagliazzo, J. 1984. Deterministic Aspects in Mathematical Demography. In press. Springer-Verlag, Berlin, Heidelberg, New York.

Keyfitz, N. 1977. Introduction to the Mathematics of Populations. With Revisions. Addison-Wesley, Reading, Massachusetts.

Levin, S. A. 1974. Dispersion and population interactions. Amer. Natur. 108:207-228.

Levin, S. A. (ed.). 1978. Populations and Communities. Studies in Mathematical Biology II. Studies in Mathematics 16. Mathematical Association of America, Washington.

Levin, S. A. 1980a. Some models for the evolution of adaptive traits. pp. 56-72 in C. Barigozzi (ed.), Vito Volterra Symposium on Mathematical Models in Biology. Springer-Verlag, Berlin, Heidelberg, New York.

Levin, S. A. 1980b. Mathematics, ecology, and ornithology. The Auk 97:422-425.

Levin, S. A. 1983a. The role of theoretical ecology in the description and understanding of populations in heterogeneous environments. Amer. Zool. 21:865-875.

Levin, S. A. 1983b. Some approaches to the modelling of coevolutionary phenomena. pp. 21-66 In M. Nitecki (ed.). Coevolution. Chicago University Press, Chicago.

Levin, S. A. and C. P. Goodyear. 1980. Analysis of an age-structured fishery model. J. Math. Biol. 9:245-274.

Lewontin, R. C. 1977. Adaptation. pp. 198-214. In Enciclopedia Einaudi Turin. 1:198-214.

Li, T. Y. and J. Yorke. 1975. Period three implies chaos. Amer. Math. Monthly. 82:985-992.

Malthus, T. R. 1978. An essay on the principle of population as it affects the future improvement of society, with remarks on the speculations of M. Godwin, M. Condorcet, and other writers. J. Johnson, London.

May, R. M. 1974. Stability and Complexity in Model Ecosystems. (Second Edition). Princeton University Press, Princeton.

May, R. M. 1978. Mathematical aspects of the dynamics of animal populations. pp. 317-366. In S. A. Levin (ed.) Populations and Communities, Mathematical Association of America, Washington.

Mayr, E. and W. Provine. 1980. The Evolutionary Synthesis. Harvard University Press, Cambridge, Massachusetts.

Maynard Smith, J. 1982. Evolution and the Theory of Games. Cambridge University Press, Cambridge.

Nagylaki, T. 1977. Selection in One- and Two-locus Systems. Springer-Verlag, Berlin, Heidelberg, New York.

Nisbet, R. M. and W. J. C. Gurney. 1982. Modelling Fluctuating Populations. Wiley, Chichester, New York.

Okubo, A. 1980. Diffusion and Ecological Problems: Mathematical Models. Springer-Verlag, Berlin, Heidelberg, New York.

Pearson, K. 1896. Contributions to the mathematical theory of evolution. Note on reproductive selection. Proc. Royal Society. 59:301-305.

Pimm, S. L. 1982. Food Webs. Chapman and Hall, London.

Provine, W. 1971. The Origins of Theoretical Population Genetics. Chicago University Press, Chicago.

Quetelet, A. L. J. 1835. Sur l'homme ét le développement de ses facultes; ou essai de physique sociale. Bachelier, Paris.

Ricker, W. E. 1954. Stock and recruitment. J. Fish. Res. Board. Can. 11:559-623.

Ricker, W. E. 1958. Handbook of computations for biological statistics of fish populations. Fish. Res. Bd. Can. Bull. 119.

Roughgarden, J. 1979. Theory of Population Genetics and Evolutionary Ecology: An Introduction. Macmillan, New York.

Sadler, M. T. 1830. The Law of Population. J. Murray, London.

Scudo, F. M. and J. R. Ziegler. 1978. The Golden Age of Theoretical Ecology: 1923-1940. Springer-Verlag, Berlin, Heidelberg, New York.

Skellam, J. G. 1951. Random dispersal in theoretical populations. Biometrika 38:196-218.

Slobodkin, L. B. 1964. The Strategy of Evolution. Amer. Sci. 52:342-357.

Smith, D. and N. Keyfitz. 1977. Mathematical Demography. Springer-Verlag, Berlin, Heidelberg, New York.

Strong, D. R., Jr. 1983. Natural variability and the manifold mechanisms of ecological communities. Amer. Natur. 122:636-660.

Verhulst, P. F. 1838. Notice sur la loi que la population súit dans son accroissement. Correspondences Mathematiques et Physiques 10:113-121.

Volterra, V. 1901-2. Sui tentutive di applicazione delle matematiche alle scienze biologiche e sociali. Ann R. Univ. Roma. pp. 3-28.

Wright, S. 1980. Genic and organismic selection. Evolution 34:825-843.

Wright, S. 1982. Character change, speciation, and the higher tax. Evolution 36:427-443.

CENTER FOR APPLIED MATHEMATICS
SECTION OF ECOLOGY AND SYSTEMATICS
ECOSYSTEMS RESEARCH CENTER
CORNELL UNIVERSITY
347 CORSON HALL
ITHACA, NEW YORK 14853

Proceedings of Symposia in Applied Mathematics
Volume 30, 1984

Population Dynamics and Demography

JAMES C. FRAUENTHAL

1. Introduction. This paper surveys some of the topics which have been of recent interest in the areas of population dynamics and demography. Generally, population dynamics is concerned with models for the interaction of species and demography with models for the growth of an age-structured population. For two reasons, this paper will focus on human demography. The models tend to be less well known than those from other areas of population biology and some of the popular techniques for dealing with age-structure are finding their way into the larger area of population biology.

Although there are numerous references throughout this paper, there is one which is so useful and comprehensive that it deserves special notice. It is an annotated collection of fifty-six of the historically most important papers in mathematical demography, edited by David Smith and Nathan Keyfitz [25].

2. Population dynamics. It is customary to count only the females of a species and to assume that males are present in sufficient numbers for reproductive purposes. In addition, populations are ordinarily viewed as closed to migration both in and out. This means that the only way to join a population is to be born and the only way to leave the population is to die. However, since there is no age structure, birth and death have the same effect, but with opposite signs. Consequently, our generic, continuous time model is of the form:

1980 *Mathematics Subject Classification.* Primary 92A15.
Key words and phrases. Demography, population dynamics.

$$\frac{dP(t)}{dt} = r[P(t)]P(t) \tag{1}$$

where $P(t)$ is the number of individuals in the population at time t, and r, which is the called the *intrinsic rate of growth,* is the excess of births over deaths as a function of the number of individuals. Note that if $P(t)$ is interpreted as a vector, the model can represent multi-species interactions. Although this is an interesting topic, it will not be pursued here (see J.C. Frauenthal [7]).

2.1 The simplest models. There are several well known models for the growth of a single species in isolation. If the intrinsic rate of growth is assumed to be a constant, $r(P)=r_0$, then the population grows (or shrinks) exponentially. This trivial model is often associated with the name of the eighteenth century English cleric T.R. Malthus [18] who was interested in the apparent ability of humans to reproduce faster than evolving technology could support. Another popular model represents the intrinsic rate of growth as a linearly decreasing function of population size, $r(P)=r_0[1-P/K]$. This model, which is usually called the *logistic equation,* possesses a globally stable equilibrium point at $P=K$, with K called the *carrying capacity of the environment.* This second model is usually associated with the name of the nineteenth century demographer, P.F. Verhulst [27]. It is worth noting that the logistic model has found wide application over the years as biologists and demographers have attempted to fit data to theory. One can, of course, go on in this manner increasing the order of the polynomial representation of the intrinsic rate of growth. However, it is really unreasonable to try to justify any form higher than second order on physical (biological) grounds.

2.2 Some alternative formulations. One immediately obvious defect of the first models discussed is that all treat population numbers as continuous and deterministic. An alternative formulation is as a birth and death stochastic process. Models of this sort are employed in so many branches of the sciences that they are not really peculiar to population modeling and will not be discussed here (see Goel and Richter-Dyn [8]). Another way to introduce non-determinism into the models is by adding noise terms directly to the differential equations to produce stochastic differential equations. Since there is not yet agreement as to the proper solution techniques, as evidenced by the Ito/Stratonovich controversy, this area has generated considerable mathematical, as well as biological, interest (see J. Roughgarden [23]).

Another obvious defect of the earlier models is that they do not allow for the effects of past population size. One can, for example, replace the logistic kernel by a term of the form

$$r(P) = r_0\left[1-\int_{-\infty}^{t} P(\tau)\,Q(t-\tau)\,d\tau\right] \tag{2}$$

where

$$\int_0^{\infty} Q(\xi)\,d\xi = 1/K$$

For $Q(\xi)$ a gamma density, this model converts the non-oscillatory stable point at $P=K$ into an oscillatory stable point (see R.M. May [20]). The lesson to be learned here is that in general the effect of time delays is destabilizing. Another interesting set of models result if $Q(\xi)$ is taken to be a Dirac delta function. This converts the ordinary differential equation into a delay differential equation (see Cushing [3]). Delay differential equations are often used as models of epidemic processes, where the delay corresponds with an incubation period.

The entire system of equations may also be recast in discrete time; the result is a system of difference equations. Equation (1) for a single species might be replaced, for example, by

$$P_{t+1} = R[P_t]P_t \tag{3}$$

where R now plays a role similar to the intrinsic rate of growth, and P_t is the population at time t. Considerable work has been done on difference equations of this form both in the mathematical and the biological literature (see T.Y. Li and J.A. Yorke, [15] and R.M. May, [19]). Typically, models of this sort are used to describe the growth of species such as insects and certain fishes which live in isolated generations. It is well known that for large ranges of the parameter values, such models admit an aperiodic trajectory which is almost indistinguishable from the trace of a random process. Such behavior has been aptly called chaos.

3. Demography. As with the models just discussed, it is assumed that only females are counted and that no migration occurs. This means that all new members of the population enter at age zero and are then subjected to mortality as they grow older. Consequently, separate functions will be needed to specify births and deaths. Before defining these, it is convenient to introduce $\rho(a,t)$, the *population age density,* which is defined such that $\rho(a,t)da$ counts the number of individuals in the population between ages a and $a+da$ at time t. It then follows that the total population, $P(t)$, is given by

$$P(t) = \int_0^{\infty} \rho(a,t)\,da. \tag{4a}$$

where $P(t)$ plays the same role here that it did in the models without age-structure. Recognizing that age and time increase at the same rate allows the basic model for the population age density to be derived easily from the formal definition of the derivative; the result is

$$\frac{\partial \rho(a,t)}{\partial a} + \frac{\partial \rho(a,t)}{\partial t} + \mu(a,P)\,\rho(a,t) = 0. \tag{4b}$$

The function $\mu(a,P)$ is called the *force of mortality*, and is defined such that $\mu(a,P)da$ is the probability of dying within the age interval from a to $a+da$ when the total population size is P. The production of newborns is assumed to be distributed over some age interval according to a *fertility* or *maternity function*, $\beta(a,P)$. This function is defined such that $\beta(a,P)da$ is the number of daughters born to a woman during the age interval a to $a+da$ when the population is P. Typically, there is an age below which fertility is zero (menarche) and and age beyond which fertility is zero (menopause). The band-limited nature of the function is useful mathematically. It follows that

$$B(t) = \rho(0,t) = \int_0^\infty \beta(a,P)\,\rho(a,t)\,da \tag{4c}$$

where $B(t)$ is called the *birth trajectory*. The set of equations (4a-c) was first derived in 1926 by the Scottish military officer and mathematician A.C. McKendrick [17]. Perhaps due to their complexity (a hyperbolic partial differential equation with two integral side conditions) these equations were overlooked until their reintroduction in 1959 in a paper by H. von Foerster [28].

3.1 Mortality and the life table. An age cohort is a group of individuals who were born in the same year. Cohorts experience the mortality which is appropriate to their age group at the time they are the particular age. This gives rise to two logical ways to report mortality experience. One is the cohort mortality which reports the mortality experience of a particular cohort as it passes through successive ages in successive years. The second is the period mortality which reports the mortality as experienced at a certain point (or short interval) of time for each cohort alive at that time.

Imagine a synthetic cohort which is composed of a fixed size initial group of newborns (typically taken to be 100,000) at a particular point in time. These individuals are subjected to the current newborn mortality experience for a year. The survivors are simultaneously subjected to the one-year-old mortality experience for a year, and so forth for all subsequent ages. Thus, during a single year the synthetic cohort is subjected to the period mortality experience of all age groups in the population. The survival statistics for the synthetic cohort is called the *life table*. To compute the life table, assume that the force of mortality is independent of secular changes. This allows (4b) to be reduced to an ordinary differential equation for $\rho(a,t) = p(a)$ which is called the *survivorship function*. The equation is easily integrated formally to yield

$$p(a+n) = p(a)\, e^{-\int_0^n \mu(a+x)\,dx} \tag{5}$$

Developing a life table from census data and vital statistics raises some interesting questions in data analysis and approximation techniques. For a discussion of a non-iterative method, see Keyfitz and Frauenthal [12].

3.2 Stable population theory. Continuing with the assumption that mortality does not change with time allows Equation (4c) to be replaced by an integral equation which is familiar from renewal theory:

$$B(t) = G(t) + \int_0^t B(t-a)\, p(a)\, \beta(a,P)\, da \tag{6}$$

where $G(t)$ accounts for births to women in the population at time zero and the integral represents births due to all successive generations of their female offspring. This formulation was first introduced in 1911 by F.R. Sharpe and A.J. Lotka [24], who subsequently noticed that a formal solution exists if one assumes that the fertility depends only on age. This formal solution, which is referred to as *stable population theory,* is discussed at length in Keyfitz [11] and Pollard [22]. It is easily shown that the solution is of the form:

$$B(t) = \sum_{i=0}^{\infty} Q_i\, e^{r_i t} \tag{7a}$$

where the r_i are the roots of the characteristic equation

$$\int_0^{\infty} e^{-r_i a}\, p(a)\, \beta(a)\, da = 1 \tag{7b}$$

and

$$Q_i = \frac{\int_0^{\infty} e^{-r_i t}\, G(t)\, dt}{\int_0^{\infty} a e^{-r_i a}\, p(a)\, \beta(a)\, da} \tag{7c}$$

It is easily shown that the characteristic equation admits one real and an infinite set of complex, conjugate root pairs, all with real parts smaller than the real root. Consequently, as time passes, the effects of the conjugate root pairs die away relative to the real root, and the birth trajectory grows exponentially. In essence, this is just an age-structured analog of the Malthusian model discussed earlier. Obviously, many extensions are possible, though few have been investigated seriously to date.

There is a discrete time version of the model just discussed which was introduced in 1945 by the British biologist P.H. Leslie [14]. This formulation has received considerable attention from both the human demography and population biology communities because of its computational convenience. Since one ordinarily keeps track of age groups instead of exact ages, a discrete formulation is useful. Further, the Leslie model simultaneously accounts for all age groups instead of just the trajectory of births.

If by analogy with the population age density, $\rho(a,t)$, a population age group density $k_{i,t}$ is defined for age groups $i=1,2,...,n$ and times steps $t=0,1,2,...$, it then follows that survival to the next age class can be modeled by the expression

$$k_{i+1,t+1} = P_i\, k_{i,t} \qquad : i=1,2,...,n \quad ; t=0,1,2,... \tag{8a}$$

where P_i is the fraction of the females in age group i at time t who survive to be in age group $i+1$ at time $t+1$. Similarly, reproduction can be represented by the expression

$$k_{1,t+1} = \sum_{i=1}^{n} M_i\, k_{i,t} \qquad : t=0,1,2,... \tag{8b}$$

where M_i is the number of daughters per female in age group i who survive through the time interval in which they are born. Clearly, these two equations can be neatly represented in matrix form. A discrete analog of stable population theory follows easily from standard matrix operations. For details, see Keyfitz [11] and Pollard [22].

3.3 Momentum of population growth. Stable population theory can be employed to determine certain properties of populations. For example, assume that a population has been growing in a stable manner for a long time, so that

$$\rho(a,t) = B_0 e^{r_0 t} \left\{ e^{-r_0 a} p(a) \right\} \tag{9}$$

A common measure of the rate of growth of the population is the *net reproduction rate*

$$R_0 = \int_0^{\infty} \beta(a) p(a) da \tag{10}$$

Note that $R_0 > 1$ corresponds with $r_0 > 0$. Next, assume that at $t=0$ there is an abrupt shift in maternity so that $\bar{R}_0 = 1$ (and thus $\bar{r}_0 = 0$.) Further, the maternity shift occurs by a uniform scaling at all ages so the new maternity function becomes $\bar{\beta}(a) = \beta(a)/R_0$. After the transients die out, $\bar{B}(t) = B_\infty$ and $\bar{\rho}(a,t) = B_\infty p(a)$. It is not hard to show that the ratio of the ultimate to the initial population size is accurately approximated by $\sqrt{R_0}$ (see Frauenthal [5]). This result demonstrates that a stably growing population which immediately assumes reproduction which eventually just balances mortality possesses a growth momentum which is due to its age structure.

3.4 Microdemography of kinship. Yet another use of stable population theory results from the observation that stable growth implies a fixed genealogy. Goldman [9] has derived methods for using a probabilistic interpretation of fertility and mortality to allow a sample survey of a population concerning living and surviving kin to be used to estimate the vital rates for the population as a whole.

3.5 Strong, weak and stochastic ergodicity. Several interesting theorems can be proven concerning the age-structure of populations subjected to fixed or changing regimes of mortality and fertility. These can generally be proven for both the continuous and the discrete formulations of the problem. In effect, stable population theory contains the strong ergodic theorem which can be stated as follows. Imagine that there are two isolated initial populations which are subject to identical, *fixed* vital rates. As time passes, the two populations asymptotically approach having the identical age structure.

The weak ergodic theorem is considerably harder to prove (see Lopez [16] for the original solution and Parlett [21] for a recent interpretation.) This theorem may be stated as follows. Imagine that there are two isolated initial populations which are subject to identical, *changing* vital rates. As time passes, the two populations asymptotically approach having the identical age structure. Recently Cohen [1,2] has extended these theorems to include the moments of the age structures of populations whose vital rates are chosen independently according to a Markov chain.

3.6 Easterlin's hypothesis. One possible extension of stable population theory is based upon an observation by Easterlin [4] that human fertility varies inversely with the cohort size of the mother. This idea was introduced into the discrete formulation by Lee [13] and the continuous formulation by Frauenthal [6]. Equation (4c) takes the form

$$B(t) = G(t) + \int_0^t B(t-a)\,p(a)\,\beta(a,t)\,da \tag{11a}$$

where

$$\beta(a,t) = \beta(a)\,M[B(t-a)]. \tag{11b}$$

If $\beta(a)$ is normalized so that

$$\int_0^\infty p(a)\,\beta(a)\,da = 1 \tag{11c}$$

then $M[B(t-a)]$ may be interpreted as the cohort net reproduction rate. This is represented in the form

$$B\,M[B] = E + (1-\gamma)(B-E) + g(B-E) \tag{11d}$$

where $B=E$ is the equilibrium birth trajectory, γ is a parameter which measures the strength of the Easterlin effect, and the function $g(B-E)$ contains only the quadratic and higher terms.

It is not hard to show by linearized analysis that there is a critical value of γ which has associated with it a doubling of the period of oscillation of the dominant roots of the solution. Recently, Swick [26] showed that for reasonable conditions on the shape of $M[B]$, a Hopf bifurcation occurs as the period doubles.

3.7 Age independent mortality. Although very little can be said in general about the basic model (4a-c), another computationally useful transformation is available if it is assumed that the force of mortality is not a function of age and that the fertility function can be represented in the form:

$$\beta(a,P) = \beta(P)\frac{a^n}{n!}e^{-\alpha a} \tag{12}$$

where n is a non-negative integer. Although the assumption of age independent mortality is severe, it may be realistic for a species living in a harsh environment. Given certain physically reasonable restrictions, M.E. Gurtin and R.C. MacCamy [10] have demonstrated that the partial differential equation (4b) can be transformed into a set of $n+2$ coupled, non-linear, ordinary differential equations of the form:

$$\frac{dA_0}{dt} = -[\mu(P)+\alpha]A_0 + \frac{1}{n!}\beta(P)A_n \tag{13a}$$

$$\frac{dA_m}{dt} = -[\mu(P)+\alpha]A_m + m\,A_{m-1} \quad : m=1,2,\ldots,n \tag{13b}$$

$$\frac{dP}{dt} = -\mu(P)P + \beta(P)A_n \tag{13c}$$

where the auxiliary variables are defined by

$$A_m(t) = \frac{1}{n!}\int_0^\infty a^m e^{-\alpha a}\rho(a,t)\,da \quad : m=0,1,\ldots,n \tag{13d}$$

and

$$B(t) = \beta(P)A_n(t). \tag{13e}$$

This set of equations is particularly attractive because it allows for simple models of age-structured population interactions analogous to those discussed earlier. Substantial amounts of work still remains to be done in this area.

4. Concluding remarks. This paper attempts to highlight topics of continuing interest in the area of population mathematics. Although large amounts of work have been done on the interaction of species, this area was mostly ignored in favor of a discussion of models which investigate the development of an age-structured population. This area seems to be attracting the interest of mathematical biologists, even though it finds most of its historical origins in the work of human demographers. Based on the nature of the problems which arise as one attempts to introduce age-structure into models of population interactions, it is safe to say that many hard and interesting practical problems still exist in this area of mathematical biology.

BIBLIOGRAPHY

1. J.E. Cohen, *Ergodicity of age structure in populations with Markovian vital rates. I. Countable states,* J. Amer. Statist. Assoc. A, **71,** (1976), 532-583.
2. J.E. Cohen, *Ergodicity of age structure in populations with Markovian vital rates. II. General states,* Adv. Appl. Prob., **9,** (1977), 18-37.
3. J.M. Cushing, *Integrodifferential Equations and Delay Models in Population Dynamics,* Springer-Verlag, New York, 1977.
4. R.A. Easterlin, *The American baby boom in historical perspective,* Amer. Econ. Rev., **51,** (1961), 869-911.
5. J.C. Frauenthal, *Birth trajectory under changing fertility conditions,* Demography, **12,** (1975), 447-454.
6. J.C. Frauenthal, *A dynamic model for human population growth,* Theo. Pop. Biol., **8,** (1975), 64-73.
7. J.C. Frauenthal, *Introduction to Population Modeling,* Birkhäuser, Boston, 1980.
8. N. Goel and N. Richter-Dyn *Stochastic Models in Biology,* Academic Press, New York, 1974.
9. N. Goldman, *Estimating the intrinsic rate of increase of a population from the average numbers of younger and older sisters,* Demography, **15,** (1978), 499-507.
10. M.E. Gurtin and R.C. MacCamy, *Some simple models for nonlinear age-dependent population dynamics,* Math. Biosci., **43,** (1979), 199-211.
11. N. Keyfitz, *Introduction to the Mathematics of Population with Revisions,* Addison-Wesley, Reading, Massachusetts, 1977.
12. N. Keyfitz and J.C. Frauenthal, *An improved life table method,* Biometrics, **31,** (1975), 889-899.
13. R.D. Lee, *The formal dynamics of controlled populations and the echo, the boom and the bust,* Demography, **11,** (1974), 563-585.
14. P.H. Leslie, *On the use of matrices in certain population mathematics,* Biometrika, **33,** (1945), 183-212.
15. T.Y. Li and J.A. Yorke, *Period three implies chaos,* Am. Math. Monthly, **82,** (1975), 985-992.
16. A. Lopez, *Weak Ergodicity,* from Problems in Stable Population Theory, Office of Population Research, Princeton, 1961.
17. A.C. McKendrick, *Applications of mathematics to medical problems,* Proc. Edinburgh Math. Soc., **44,** (1926), 98-130.
18. T.R. Malthus, *An Essay on the Principle of Population,* Printed for J. Johnson in St. Paul's Churchyard, London, 1798.
19. R.M. May, *Simple mathematical models with very complicated dynamics,* Nature, **261,** (1976), 459-467.
20. R.M. May, *Stability and Complexity in Model Ecosystems,* Princeton University Press, Princeton, 1973.

21. B. Parlett, *Ergodic Properties of Populations I: The One Sex Model,* Theo. Pop. Biol., **1,** (1970), 191-207.
22. J.H. Pollard, *Mathematical Models for the Growth of Human Populations,* Cambridge University Press, Cambridge, 1973.
23. J. Roughgarden, *Theory of Population Genetics and Evolutionary Ecology: An Introduction,* Macmillan, New York, 1979.
24. F.R. Sharpe and A.J. Lotka, *A problem in age distribution,* Phil. Mag., **21,** (1911), 435-438.
25. D. Smith and N. Keyfitz (eds.), *Mathematical Demography,* Biomathematics Vol. 6, Springer-Verlag, New York, 1977.
26. K.E. Swick, *A nonlinear model for human population dynamics,* SIAM J. Appl. Math., **40,** (1981), 266-278.
27. P.F. Verhulst, *Notice sur la loi que la population suit dans son accroissement,* Correspondance mathématique et physique publée par A. Quételet (Brussels), **X** (1838), 113-121.
28. H. von Foerster, *Some remarks on changing populations,* in The Kinetics of Cellular Proliferation, Grune and Stratton, New York, 1959.

AT&T-BELL LABORATORIES, HOLMDEL, NEW JERSEY 07733

Proceedings of Symposia in Applied Mathematics
Volume 30, 1984

SOME MATHEMATICAL PROBLEMS IN POPULATION GENETICS

Thomas Nagylaki[1]

ABSTRACT. Two areas of theoretical population genetics, geographical variation and the evolutionary effects of molecular turnover processes, are discussed, with emphasis on unsolved problems. The deterministic problems primarily involve functional iteration; the stochastic ones concern Markov chains and diffusion processes. In the area of geographical variation, limiting results, invariance principles, and robustness are treated. Among molecular turnover processes, most attention is devoted to gene conversion at a single locus.

1. INTRODUCTION

Population genetics concerns the genetic structure and evolution of natural populations. The genetic composition of a population is usually described by genotypic proportions, which may depend on space and time. These genotypic frequencies are determined by a few elementary genetic principles and the following evolutionary factors.

Various genotypes may have different probabilities of surviving to adulthood and may reproduce at different rates. Differential mortality and fertility are the components of *selection.* Unless the population is in equilibrium, selection will change the genotypic and allelic frequencies in accordance with the expected number of progeny, called fitness, of the various genotypes. Natural selection has been recognized since Darwin as the directive force of adaptive evolution.

The action of selection is strongly affected by the *mating system.* If mating occurs without regard to the genotypes under consideration, we say it is random. This is the simplest situation and, at least approximately, appears to be frequently realized in nature. We say there is inbreeding if related individuals are more likely to mate than randomly chosen ones. Assortative mating refers to the tendency of individuals who are similar with respect to the trait

1980 *Mathematics Subject Classification.* 92A10.
Key words and phrases. Geographical variation, subdivided populations, stepping-stone model, gene conversion.
[1]Supported by National Science Foundation Grant DEB81-03530.

in question to mate with each other. Disassortative mating means that phenotypically dissimilar individuals mate more often than randomly chosen ones. Nonrandom mating influences genotypic frequencies. In the absence of selection, inbreeding does not change gene frequencies, but assortative and disassortative mating may. This may happen if the mating pattern is such that some genotypes have a higher probability of mating than others.

Mutation designates the change from one allelic form to another. Clearly, it directly alters gene frequencies.

In spatially structured populations, *migration* must be taken into account. It can affect not only the geographical composition of the population, but the amount of genetic variability as well.

Unless some of the parameters, such as the selection intensities, required to specify the elements of evolution described above fluctuate at random, these evolutionary forces will be deterministic. In a finite population, however, allelic frequencies will vary probabilistically due to the random sampling of genes from one generation to the next. This process is called *random genetic drift*. Its causes are (nonselective) random variation in the number of offspring of different individuals and the stochastic nature of Mendel's Law of Segregation. Evidently, the smaller the population, the larger is the evolutionary role of random drift. No matter how large the population is, however, the fate of rare genes still depends strongly on random sampling.

The genomes of higher organisms contain a significant proportion of repeated DNA sequences, which may be arranged tandemly on a chromosome or dispersed throughout one or more chromosomes. Within a single phylogenetic species, the DNA sequences within one family of repeated genes are often very similar. Multigene families that serve a similar function in closely related species, however, may show considerable divergence from each other (Arnheim, 1983). These observations have stimulated the investigation of the evolutionary effects of *molecular turnover processes*: unequal crossing over, replicative transposition, and gene conversion.

Unequal crossing over between sister chromatids can lead to sequence homogeneity of tandem arrays (Smith, 1973; Tartof, 1973). In a finite population, the same holds for unequal crossing over between homologous chromosomes. In Figure 1, we display schematically the effect of unequal crossing over; A_1, A_2, and A_3 represent repeats. Ohta (1983a and references therein) has studied extensively various evolutionary models of this process. Although the simplest case (Ohta, 1976) has been formulated and solved exactly (Nagylaki and Petes, 1982, p. 332), challenging mathematical problems remain in more general and important situations.

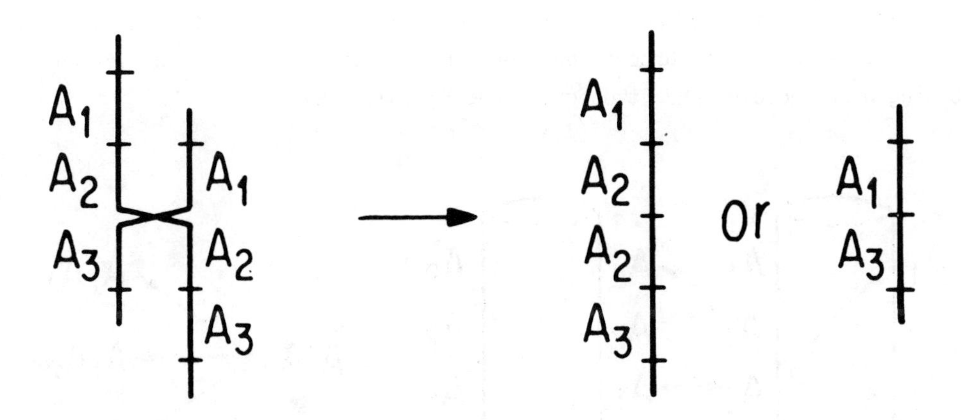

FIGURE 1. Unequal crossing over.

In the genomes of both lower and higher organisms are found DNA sequences that are capable of replication and transposition to new sites (Campbell, 1983). The distribution in a population of the number of such transposable elements per individual is of great biological interest and has been recently investigated (Charlesworth and Charlesworth, 1983; Kaplan and Brookfield, 1983; Langley, Brookfield, and Kaplan, 1983). Further research is required in this area, especially with regard to the effects of natural selection, linkage, and finiteness of the number of available sites (Charlesworth and Charlesworth, 1983).

Gene conversion is the non-reciprocal transfer of information between two genes (Nagylaki and Petes, 1982). As indicated in Figure 2, this can occur between genes on a single chromatid, on sister chromatids, and on different (homologous or non-homologous) chromosomes. Gene conversion can produce sequence homogeneity of tandem and dispersed multigene families (Edelman and Gally, 1973) and can significantly influence evolution at individual loci (Gutz and Leslie, 1976; Lamb and Helmi, 1982; Nagylaki, 1983a,b; Walsh, 1983). Conversion is biased if one of two interacting genes has a higher probability of converting the other than *vice versa*. The dynamics of the probabilities of genetic identity under unbiased gene conversion has been examined at the population level (Ohta, 1982, 1983a,b; Nagylaki, 1983c; Ohta and Dover, 1983). A more complete analysis of this stochastic process has not been performed. The major open problem is the extension of the study of biased conversion in a single chromosome lineage (Nagylaki and Petes, 1983) to a population. Note also that the creation of new alleles by gene conversion has not been explored (Nagylaki, 1983a). This process has been investigated, however, for intragenic crossing over (Golding and Strobeck, 1983; Hudson, 1983).

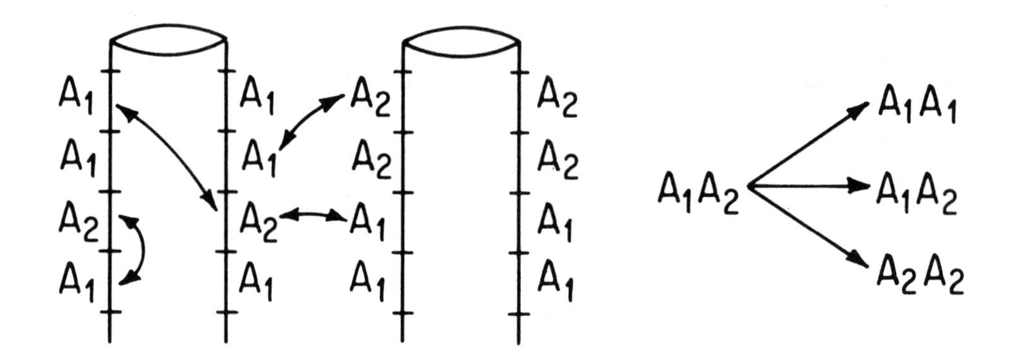

FIGURE 2. Gene conversion.

The analysis of the joint action of all the above evolutionary forces is prohibitively difficult, and even if it were possible, would almost certainly lead to results too complex to be illuminating. We can obtain insight by analyzing models that incorporate judiciously chosen subsets of evolutionary forces. In Sections 2 and 3, we treat some aspects of geographical variation and gene conversion at a single locus, respectively. We present the biological assumptions and the results; for discussions and proofs, the reader should consult the references.

The genetic background required is very briefly sketched and genetic terms are indexed and defined in Nagylaki (1977a); Crow (1983) expounds basic genetics lucidly and concisely. Malécot (1948), Kempthorne (1957), Moran (1962), Ewens (1969, 1979), Wright (1969), Crow and Kimura (1970), Maruyama (1977), Nagylaki (1977a), Bulmer (1980), and Kingman (1980) treat various aspects of theoretical population genetics. The bibliography of Felsenstein (1981) has almost complete coverage of the field through the end of 1980.

2. GEOGRAPHICAL VARIATION

To achieve some coherence of biological and mathematical formulation, several topics properly subsumed under the title of this section will be omitted.

The discrete models mentioned in this paragraph involve multidimensional non-linear functional iteration; the continuous ones concern nonlinear parabolic partial differential equations. Nagylaki (1977a, Ch. 6) and Karlin (1982) treat migration-selection polymorphism in subdivided populations. There is also a rich literature on the existence, uniqueness, and stability of such polymorphisms in spatially and temporally continuous models (Fife and Peletier, 1981; Yanagida, 1982). The rigorous incorporation of random genetic drift into these models is difficult (Nagylaki and Lucier, 1980), and many basic problems are unsolved in this area. Weinberger (1982) discusses the wave of advance of genes favored uniformly throughout the habitat. Finally, genotype-dependent migration leads to surprisingly hard mathematics even in the absence of all other evolutionary forces (Nagylaki and Moody, 1980; Moody, 1981); here too, major problems of existence, uniqueness, and stability remain open.

Unless specifically noted otherwise, the following assumptions will hold throughout this section.

(i) The population is diploid and monoecious.

(ii) A finite number of randomly mating colonies exchange migrants.

(iii) Generations are discrete and nonoverlapping.

(iv) The analysis is restricted to a single locus.

(v) The migration pattern is fixed and ergodic (i.e., time independent, irreducible, and aperiodic).

Let N_i represent the number of adults in colony i. We employ the backward migration matrix, M, to describe the pattern of dispersion: m_{ij} designates the probability that an individual or gamete in colony i comes from colony j. Migration is *conservative* if and only if it does not change the subpopulation numbers. By the ergodicity of the nonnegative stochastic matrix M, the eigenvalue one of M is nondegenerate (= simple) and exceeds all other eigenvalues in modulus; we may choose the left eigenvector ν corresponding to this unit eigenvalue to have only positive components (Gantmacher, 1959, Vol. II, Ch. 13). Thus, the conditions

$$0 < \nu_i < 1, \qquad \sum_i \nu_i = 1, \qquad \sum_j \nu_j m_{ji} = \nu_i \tag{1}$$

uniquely determine ν. Let κ_i and N_T denote the proportion of adults in colony i and the total population number. Then

$$\kappa_i = N_i/N_T, \qquad 0 < \kappa_i < 1, \qquad \sum_i \kappa_i = 1. \tag{2}$$

The effective population number

$$N_e = \beta N_T, \qquad \beta = \left(\sum_i \nu_i^2/\kappa_i \right)^{-1}, \tag{3}$$

satisfies $N_e \leqslant N_T$, with equality if and only if $\kappa = \nu$, which occurs if migration is conservative (Nagylaki, 1980).

We treat neutral and selective models separately.

2.1 *Neutral Models*

We focus here on general migration patterns. Sawyer (1976, 1977a,b, 1978, 1979), Rusinek (1982), and Sawyer and Felsenstein (1983) analyze some important special cases in depth; see Nagylaki (1977b, 1978) for references to the rest of this extensive literature. In this subsection, we impose two more assumptions.

(vi) There is no selection at the locus under consideration.
(vii) Every allele mutates at rate u $(0 \leqslant u \leqslant 1)$ per generation to new alleles.

We examine two models: gametic and diploid dispersion. In both of them, the subpopulation numbers are finite only at the adult stage and random drift operates through population regulation.

For gametic dispersion, let $f_{ij}(t)$ designate the probability that two distinct genes chosen at random from adults just before reproduction in generation t (= 0,1,2,...), one from colony i and one from colony j, are the same allele. The probability of allelic identity is (in principle) a measurable functional of our rather complicated finite Markov chain. Its complement, $1 - f_{ij}$, is an index of the amount and pattern of genetic diversity in the population. In particular, $1 - f_{ii}$ is the expected heterozygosity in deme i. We summarize our model in the life cycle below.

$$\underset{N_i, f_{ij}}{\text{Adults}} \xrightarrow[\text{gametogenesis}]{} \underset{\infty, -}{\text{Gametes}} \xrightarrow[\text{dispersion}]{} \underset{\infty, -}{\text{Gametes}} \xrightarrow[\text{fertilization}]{}$$

$$\underset{\infty, -}{\text{Zygotes}} \xrightarrow[\text{mutation}]{} \underset{\infty, -}{\text{Zygotes}} \xrightarrow[\text{regulation}]{} \underset{N_i, f'_{ij}}{\text{Adults}}$$

Then (Malécot, 1951; Sawyer, 1976; Nagylaki, 1980)

$$f'_{ij} = v\left[\sum_{kl} m_{ik} m_{jl} f_{kl} + \sum_{k} m_{ik} m_{jk} (2N_k)^{-1} (1 - f_{kk})\right] , \tag{4}$$

where $v = (1 - u)^2$ and the prime indicates the next generation. Complete random union of gametes implies that a proportion $1/N_i$ of the individuals in colony i are produced by self-fertilization.

As $t \to \infty$, $f_{ij}(t) \to \hat{f}_{ij}$, the unique equilibrium of (4); $\hat{f}_{ij} = 1$ for every i and j, which corresponds to a completely homogeneous population, if and only if $u = 0$, i.e., there is no mutation (Malécot, 1951; Nagylaki, 1980).

In contrast to the above model, suppose now that diploids disperse and selfing does not occur. [There is a similar theory for selfing at rate $1/N_i$

(Nagylaki, 1983d)]. Let $I_{ij}(t)$ represent the probability that two genes chosen at random from distinct adults just before reproduction in generation t, one from deme i and one from deme j, are the same allele. We designate by $J_i(t)$ the probability that the two genes of an adult chosen at random from deme i just before reproduction in generation t are the same allele. Thus, J_i is the expected homozygosity in colony i. We display our formal scheme below.

$$\underset{N_i, I_{ij}, J_i}{\text{Adults}} \xrightarrow[\text{reproduction}]{} \underset{\infty,-,-}{\text{Zygotes}} \xrightarrow[\text{migration}]{} \underset{\infty,-,-}{\text{Zygotes}} \xrightarrow[\text{mutation}]{}$$

$$\underset{\infty,-,-}{\text{Zygotes}} \xrightarrow[\text{regulation}]{} \underset{N_i, I'_{ij}, J'_i}{\text{Adults}}$$

Instead of (4), now we have the more complicated model (Sawyer, 1976; Nagylaki, 1983d).

$$I'_{ij} = v\left[\sum_{kl} m_{ik}m_{jl}I_{kl} + \sum_{k} m_{ik}m_{jk}(2N_k)^{-1}(1 + J_k - 2I_{kk})\right], \tag{5a}$$

$$J'_i = v\sum_k m_{ik}I_{kk}. \tag{5b}$$

As $t \to \infty$, $[I_{ij}(t), J_i(t)] \to [\hat{I}_{ij}, \hat{J}_i]$, the unique equilibrium of (5); $\hat{I}_{ij} = 1$ and $\hat{J}_i = 1$ for every i and j if and only if $u = 0$ (Sawyer, 1976; Nagylaki, 1983d).

In the remainder of this subsection, we investigate three aspects of the models (4) and (5). Since the detailed analysis of (4) and (5) is difficult and often a geographically structured population behaves genetically like a panmictic one, it is natural to seek properties that are invariant under population subdivision. Even if the property being examined is not exactly invariant, approximate invariance may hold if migration is much stronger than all the other evolutionary forces, an idea that leads us to study the strong-migration limit. Finally, it is important to ask how well the classical model (4) approximates the usually more realistic model (5); this is an explicit and necessary, but not sufficient, criterion for robustness.

2.1a Invariance

Maruyama's pioneering work on this problem is discussed in Nagylaki (1982). Our first two results hold for gametic dispersion.

For $u > 0$, the weighted means

$$\bar{f} = \sum_{ij} \nu_i \nu_j f_{ij}, \qquad \bar{f}_0 = \beta \sum_i (\nu_i^2/\kappa_i) f_{ii} \tag{6}$$

satisfy (Nagylaki, 1982)

$$\hat{\bar{f}} = \frac{v(1 - \hat{\bar{f}}_0)}{2N_e(1 - v)} \approx \frac{1 - \hat{\bar{f}}_0}{4N_e u} \tag{7}$$

at equilibrium, where the approximation holds if $u << 1$. Since generally u is of order 10^{-5} or less, this approximation is extremely accurate. Population subdivision influences (7) only through the stationary distribution, ν, of M and the niche proportions κ. For conservative migration, $\bar{f}$ and $\bar{f}_0$ simplify to the general and local means for genes chosen at random and $N_e = N_T$.

If $u = 0$, our stochastic process is an absorbing Markov chain. Therefore, some allele will be fixed in finite time with probability one, and hence the total number of heterozygotes formed by a new mutant (i.e., an allele initially represented exactly once in the population) is finite with probability one. For conservative migration, the expectation of this random variable is simply $2N_T$ (Maruyama, 1971; Nagylaki, 1982), as for random mating.

If $u = 0$ and migration is conservative, for both gametic and diploid dispersion a martingale argument establishes that the fixation probability of an allele is precisely its initial frequency in the entire population (Nagylaki, 1980), again as for panmixia.

See Nagylaki (1980, 1982) for further results.

2.1b The Strong-Migration Limit

To investigate (4) and (5) at equilibrium, we assume that $N_e = \mu/u$ and let $u \to 0$ with M, κ, and μ fixed. In this limit, mutation and random drift are much weaker than migration. We find (Nagylaki, 1980, 1983d)

$$\hat{f}_{ij}, \hat{I}_{ij}, \hat{J}_i \to 1/(1 + 4\mu) \tag{8}$$

as $u \to 0$.

To derive the rate and pattern of convergence to the equilibrium (8), we suppose that M is fixed and all the demes are large, $N_i \to \infty$ for every i. Thus, migration dominates random drift, but the intensity of mutation is arbitrary. By calculating the leading eigenvalue and eigenvector of the relevant linear systems, we deduce (Nagylaki, 1980, 1983d)

$$f_{ij}(t) - \hat{f}_{ij} \sim \alpha[1 + O(N_e^{-1})]v^t[1 - (2N_e)^{-1} + O(N_e^{-2})]^t, \tag{9a}$$

$$I_{ij}(t) - \hat{I}_{ij} \sim \gamma[1 + O(N_e^{-1})]v^t[1 - (2N_e)^{-1} + O(N_e^{-2})]^t, \tag{9b}$$

$$J_i(t) - \hat{J}_i \sim v\gamma[1 + O(N_e^{-1})]v^t[1 - (2N_e)^{-1} + O(N_e^{-2})]^t \tag{9c}$$

as $t \to \infty$. The unevaluated constants α and γ may not be equal; they do *not* depend on the subscripts i and j. The error terms in (9) presumably differ from one another and depend on i and j.

The only influence of population structure on (8) and (9) is the appearance of N_e instead of N_T, and this effect disappears if migration is conservative.

The reader may wonder what happens in the *weak-migration limit*. The most interesting case is that of equilibrium with $u > 0$. In view of the robustness results below, it suffices to examine (4). Put

$$m_{ij} = \delta_{ij} + m\lambda_{ij},$$

where δ_{ij} represents the Kronecker delta, $m \to 0+$, and λ_{ij} is of order one. Clearly, $\lambda_{ii} < 0$, $\lambda_{ij} \geqslant 0$ if $i \neq j$, and

$$\sum_j \lambda_{ij} = 0.$$

Substitute

$$\hat{f}_{ij} = \hat{f}_{ij}^{(0)} + m\hat{f}_{ij}^{(1)} + O(m^2)$$

into (4) at equilibrium and expand in powers of m. This leads to the zeroth-order solution, which corresponds to isolated colonies,

$$\hat{f}_{ii}^{(0)} = \frac{v}{v + 2N_i(1-v)} \approx \frac{1}{1 + 4N_i u}$$

$$\hat{f}_{ij}^{(0)} = 0, \qquad i \neq j$$

(Malécot, 1946, 1948; Kimura and Crow, 1964), and the first-order correction

$$\hat{f}_{ii}^{(1)} = \frac{4N_i\lambda_{ii}\hat{f}_{ii}^{(0)}}{v + 2N_i(1-v)} \approx \frac{4N_i\lambda_{ii}\hat{f}_{ii}^{(0)}}{1 + 4N_i u},$$

$$\hat{f}_{ij}^{(1)} = \frac{\lambda_{ji}\hat{f}_{ii}^{(0)} + \lambda_{ij}\hat{f}_{jj}^{(0)}}{1 - v} \approx \frac{\lambda_{ji}\hat{f}_{ii}^{(0)} + \lambda_{ij}\hat{f}_{jj}^{(0)}}{2u}, \qquad i \neq j,$$

where the approximations hold for $u \ll 1$. Notice that, as expected intuitively, the homogenizing effect of random drift within colonies is weakened ($\hat{f}_{ii}^{(1)} < 0$) and different colonies become genetically similar ($\hat{f}_{ij}^{(1)} > 0$ for $i \neq j$, provided $\lambda_{ij} \neq 0$ or $\lambda_{ji} \neq 0$).

2.1c Robustness

The results (8) and (9) may be viewed as robustness theorems for strong migration. For arbitrary values of the parameters, we confine ourselves to equilibrium. Therefore, to avoid trivialities, we take $0 < u < 1$. We posit that the subpopulation numbers are all the same, $N_i = N$ for every i, and migration depends only on displacement, $m_{ij} = m_{i-j}$, rather than on initial and final positions separately. For the three migration patterns considered below, these assumptions imply homogeneity of the probabilities of identity:

$$\hat{f}_{ij} = \hat{f}_{i-j}, \qquad \hat{I}_{ij} = \hat{I}_{i-j}, \qquad \hat{J}_i = \hat{J}. \tag{10}$$

With the aid of (10), we can demonstrate that

$$|\hat{J} - \hat{f}_0| < 2u + |\hat{I}_0 - \hat{f}_0|. \tag{11}$$

As discussed below (7), $u \ll 1$; therefore, by dint of (11), it suffices to establish that $\hat{I}_i$ is well approximated by $\hat{f}_i$ for every i.

For an infinite lattice in d dimensions and for colonies in a "circle,"

$$0 \leqslant \frac{\hat{f}_i}{\hat{I}_i} - 1 < \min\left(2u, \frac{1}{2N}\right) \tag{12}$$

for every i. In the case of the lattice, the number of colonies is, of course, infinite, and if $d \geqslant 2$, i represents a vector with integral components. For n equivalent islands, the elements of the backward migration matrix are given by

$$m_{ij} = \begin{cases} 1 - m, & i = j, \\ m/(n - 1), & i \neq j, \end{cases}$$

where m denotes the migration rate. Then we have

$$0 \leqslant \frac{\hat{f}_i}{\hat{I}_i} - 1 < \min\left(6u, \frac{3}{N}\right) \tag{13}$$

for every i.

Since $u \ll 1$ and generally $N \gg 1$, the bounds (12) and (13) are very stringent. Equations (10) to (13) and a number of other results are proved in Nagylaki (1983d). The proofs of (12) and (13) depend on explicit representations of the solutions. If a bound similar to (12) and (13) exists for arbitrary M, it will probably be necessary to prove it by less obvious manipulation of (4) and (5).

2.2 *Selective Models*

Since selection is intrinsically nonlinear, its treatment in a subdivided population is usually difficult. We replace assumptions (vi) and (vii) by the following two hypotheses.

(vi*) There are n $(< \infty)$ alleles, $A_1, A_2, \ldots, A_n$.

(vii*) Selection acts only through viability differences; the genotype A_jA_k has constant viability $w_{i,jk}$ in colony i.

Let the random variable $p_{i,j}$ designate the frequency of A_j in colony i in adults just before reproduction. Our life cycle is the following.

$$\underset{N_i, p_{i,j}}{\text{Adults}} \xrightarrow[\text{reproduction}]{} \underset{\infty, p_{i,j}}{\text{Zygotes}} \xrightarrow[\text{selection}]{} \underset{\infty, p^*_{i,j}}{\text{Adults}} \xrightarrow[\text{migration}]{}$$

$$\underset{\infty, p^{**}_{i,j}}{\text{Adults}} \xrightarrow[\text{mutation}]{} \underset{\infty, p^{***}_{i,j}}{\text{Adults}} \xrightarrow[\text{regulation}]{} \underset{N_i, p'_{i,j}}{\text{Adults}}$$

In this scheme, population regulation causes multinomial sampling of genotypes at the end of every generation. In the diffusion approximation, to which we restrict ourselves, this is the same as the biologically less rigorous, classical model of multinomial sampling of genes (Nagylaki, 1980, 1982). Furthermore, if, instead of adult migration, we posit juvenile migration (i.e., we interchange selection and migration in our life cycle) or gametic dispersion, the same results hold (Nagylaki, 1980, 1982), which makes them robust in the sense used above. Consequently, we need to investigate only invariance and the strong-migration limit.

2.2a Invariance

Solely for the analysis of invariance, we impose the following additional assumptions. It would be highly desirable to establish results under more general conditions.

(viii) There are two alleles, A_1 and A_2.

(ix) There is no mutation.

(x) The selection pattern is the same in all demes and is purely additive:

$$w_{11} = 1 + s, \qquad w_{12} = 1, \qquad w_{22} = 1 - s. \tag{14}$$

(The choice $w_{12} = 1$ is a convention; (14) is the most general deme-independent scheme without dominance.)

(xi) Migration is conservative.

To obtain the diffusion limit of our Markov chain, put

$$s = \sigma(2N_T)^{-1}, \qquad m_{ij} = \delta_{ij} + \mu_{ij}(2N_T)^{-1} + O(N_T^{-2}) \tag{15}$$

and let $N_T \to \infty$ with κ_i, σ, and μ_{ij} fixed for every i and j. This means that all evolutionary forces are weak, as is often the case. In this limit, the fixation probability of A_1, all moments of the total number of heterozygotes that appear in the population, these moments conditioned on fixation of A_1, and the last two functionals restricted to an infinitesimal gene frequency interval are the same as for random mating (Maruyama, 1972; Nagylaki, 1982).

2.2b The Strong-Migration Limit

Suppose A_i mutates to A_j at rate u_{ij} (by convention, $u_{ii} = 0$). We take the diffusion limit by setting

$$u_{ij} = \mu_{ij}(2N_e)^{-1}, \qquad w_{i,jk} = 1 + \sigma_{i,jk}(2N_e)^{-1} \tag{16}$$

and letting $N_e \to \infty$ with κ_i, μ_{ij}, $\sigma_{i,jk}$, and M fixed for every i, j, and k. Define the weighted mean gene frequencies P_j and the local deviations $q_{i,j}$:

$$P_j = \sum_i \nu_i p_{i,j}, \qquad q_{i,j} = p_{i,j} - P_j. \tag{17}$$

Then, as $N_e \to \infty$, $q([2N_e\tau]) \to 0$ in probability for $\tau > 0$ and $P([2N_e\tau])$ has the same limiting diffusion as for random mating, with scaled selection intensities

$$\rho_{jk} = \sum_i \nu_i \sigma_{i,jk} \tag{18}$$

(Nagylaki, 1980). Again, population subdivision manifests itself only through N_e and ν, and hence its effect disappears for conservative migration.

Assuming that there are only two alleles, the migration matrix is symmetric, and the deme sizes and selection coefficients are the same in all colonies, Slatkin (1981) has calculated the fixation probability in the weak-migration approximation. More work is needed on this topic.

3. GENE CONVERSION

The results presented in this section, which is restricted to pure gene conversion, are proved in Nagylaki (1983a); see Nagylaki (1983a,b) for analyses of the joint action of gene conversion and other evolutionary forces. We posit the following.

(i) The diploid, monoecious population mates at random.

(ii) Generations are discrete and nonoverlapping.

(iii) The analysis is confined to a single locus.

(iv) Selection, mutation, and random drift are absent. This requires that the population be very large.

(v) Even if gene conversion occurs, an A_iA_j individual can produce only A_i and A_j gametes.

A skew-symmetric matrix B determines the effect of gene conversion on allelic frequencies; we denote the probability that an A_iA_j individual produces an A_i gamete by $\frac{1}{2}(1 + b_{ij})$, where $b_{ij} = -b_{ji}$ and $|b_{ij}| \leqslant 1$ for every i and j. The allele A_i has a conversional advantage over A_j if $b_{ij} > 0$; conversion is unbiased if $b_{ij} = 0$. Let $p_i(t)$ represent the frequency of A_i in generation t $(= 0,1,2,\ldots)$. For n alleles, the single-generation changes in the allelic frequencies read

$$\Delta p_i = p_i \sum_{j=1}^{n} b_{ij} p_j, \tag{19}$$

which maps the simplex

$$p_i \geqslant 0, \qquad \sum_{i=1}^{n} p_i = 1 \tag{20}$$

into itself.

The interior, or completely polymorphic, equilibria (i.e., those with $p_i > 0$ for every i) satisfy $Bp = 0$. Therefore, if n is odd, interior equilibria *may* exist; generically (i.e., if B has rank $n-1$) there exists at most one such equilibrium. If n is even, generically $\det B \neq 0$, and then an interior equilibrium does not exist.

To study the dynamics of (19), it is fruitful to associate with each n-allelic conversion pattern an oriented graph (Harary, 1969) with n points. The point i corresponds to the allele A_i; the points i and j are connected if and only if conversion between A_i and A_j is biased ($b_{ij} \neq 0$); if A_i has a conversional advantage with respect to A_j ($b_{ij} > 0$), the directed line that connects i and j points from j to i. (See Figures 3, 4, and 5 for examples.) If the alleles can be divided into two or more disjoint sets such that conversion is biased only within each set, (19) easily reveals that these sets evolve independently of each other and the total gene frequency of each set is constant. Therefore, without loss of generality, we may restrict our attention to connected graphs.

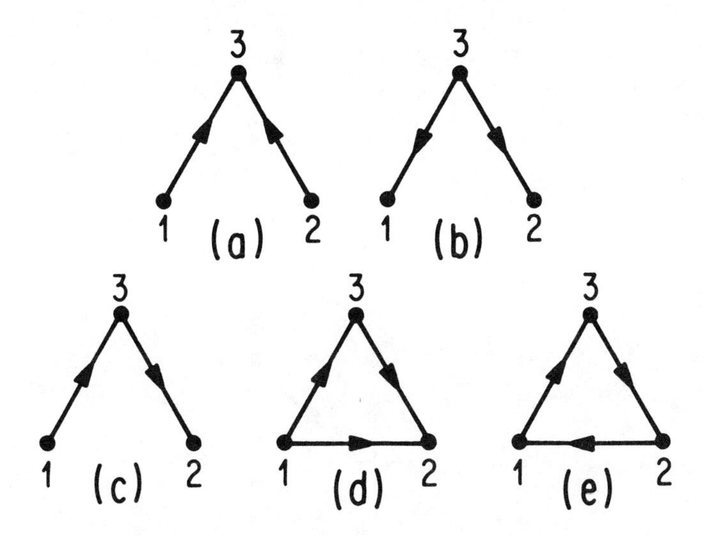

FIGURE 3. The connected oriented graphs with three points.

Next, we state three principles that enable us to analyze directly many conversion patterns.

Principle 1. If A_i has a conversional advantage relative to every allele to which it is connected (i.e., A_i is a sink), then all those alleles are ultimately lost. In particular, if a sink A_i is connected to every other allele, then it is ultimately fixed.

For instance, in Figure 4, A_1 is a sink, so $p_2, p_3, p_4 \to 0$, but $p_1 \to 1$ may not occur (e.g., Figure 3b).

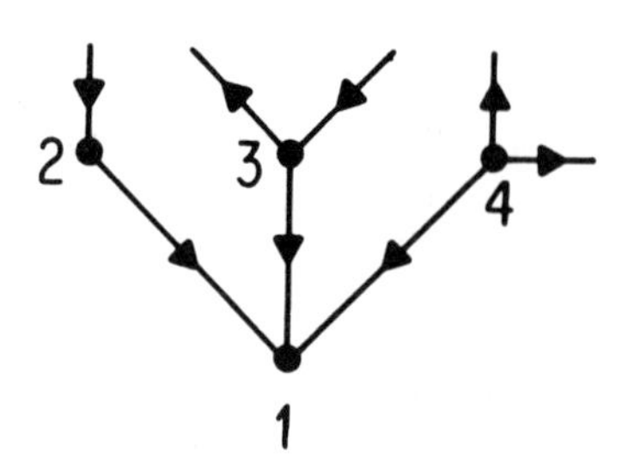

FIGURE 4. A_1 is a sink.

Principle 2. If A_i has a conversional disadvantage relative to every allele to which it is connected (i.e., A_i is a source), then A_i is lost or those alleles are lost (or possibly both). In particular, if a source A_i is connected to every other allele, then A_i is lost.

For instance, in Figure 5, A_1 is a source, so $p_1 \to 0$ or $p_2, p_3, p_4 \to 0$; $p_1 \to 0$ may not occur (e.g., Figure 3c).

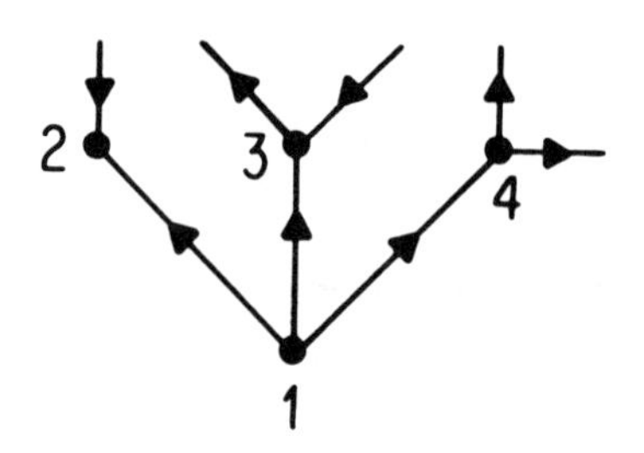

FIGURE 5. A_1 is a source.

Principle 3. If there exists a unique, isolated completely polymorphic equilibrium $\hat{p}$ and the initial gene frequency $p(0) \neq \hat{p}$, then $p(t)$ converges to the boundary of the simplex (20).

Remarks. Equivalently, we assert that $p(t)$ permanently leaves every compact interior set of the simplex. This implies

$$\liminf_{t \to \infty} p_i(t) = 0 \tag{21}$$

for some i. Notice that no allelic frequency need *remain* small: in many cases, as in Figure 6, $\lim_{t\to\infty} p_i(t) \neq 0$ for any i. From Principle 3 and the discussion at the beginning of this section of the existence of completely polymorphic equilibria, we conclude that convergence to the boundary occurs for many conversion patterns if the number of alleles is odd.

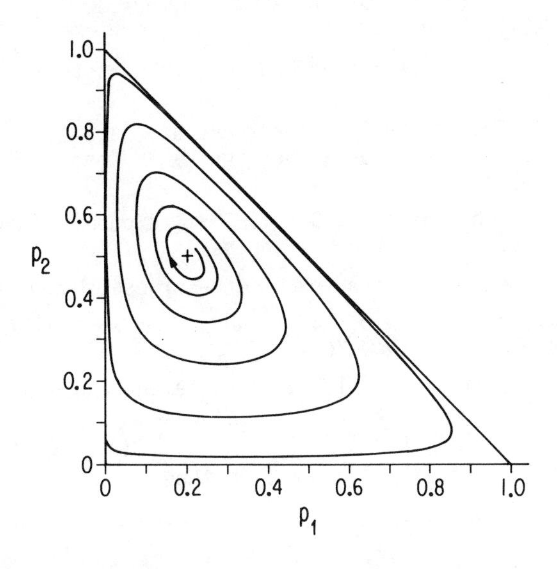

FIGURE 6. An orbit corresponding to Figure 3e.

Principles 1, 2, and 3 yield a complete analysis for $n = 2$ and $n = 3$; extensive numerical investigation of (20) was carried out for $n = 4$ and $n = 5$. The results strongly suggest the following conjectures.

Conjectures. Suppose that $p(0)$ is not an equilibrium. (*i*) If conversion is biased within at least one pair of alleles, then $p(t)$ converges to the boundary of the simplex (20). This implies that the frequency of at least one allele must become arbitrary small, i.e., (21) holds for at least one i for every connected oriented graph. (*ii*) If conversion is biased within every pair of alleles, then the frequency of at most one allele can fail to become arbitrarily small, i.e., (21) fails for at most one i for every tournament.

If these conjectures are correct, then in the absence of mutation and selection, the biological ubiquity of small random perturbations implies that biased gene conversion at a single multiallelic locus diminishes intrapopulation genetic variability by decreasing the number of alleles. If conversion is biased within every pair of alleles, ultimately only one allele survives. Which alleles are lost often depends on random genetic drift and other stochastic perturbations that may affect gene frequencies. Therefore, biased conversion must frequently contribute to the genetic divergence of isolated populations.

ACKNOWLEDGMENT

I thank T. Petes for helpful discussions.

REFERENCES

Arnheim, N. 1983. Concerted evolution of multigene families. In *Evolution of Genes and Proteins* (M. Nei and R. K. Koehn, eds.). Pp. 38-61. Sinauer, Sunderland, Mass.

Bulmer, M. G. 1980. *The Mathematical Theory of Quantitative Genetics.* Clarendon Press, Oxford.

Campbell, A. 1983. Transposons and their evolutionary significance. In *Evolution of Genes and Proteins* (M. Nei and R. K. Koehn, eds.). Pp. 258-279. Sinauer, Sunderland, Mass.

Charlesworth, B., and Charlesworth, D. 1983. The population dynamics of transposable elements. *Genet. Res.*, in press.

Crow, J. F. 1983. *Genetics Notes.* 8th ed. Burgess, Minneapolis.

Crow, J. F., and Kimura, M. 1970. *An Introduction to Population Genetics Theory.* Harper and Row, New York.

Edelman, G. M., and Gally, J. A. 1970. Arrangement and evolution of eukaryotic genes. In *Neurosciences: Second Study Program* (F. O. Schmitt, ed.). Pp. 962-972. Rockefeller University Press, New York.

Ewens, W. J. 1969. *Population Genetics.* Methuen, London.

Ewens, W. J. 1979. *Mathematical Population Genetics.* Springer, Berlin.

Felsenstein, J. 1981. *Bibliography of Theoretical Population Genetics.* Dowden, Hutchinson, and Ross, Stroudsburg, Pa.

Fife, P. C., and Peletier, L. A. 1981. Clines induced by variable selection and migration. *Proc. Roy. Soc. Lond.* B **214**, 99-123.

Gantmacher, F. R. 1959. *The Theory of Matrices.* Vol. II. Chelsea, New York.

Golding, G. B., and Strobeck, C. 1983. Increased number of alleles found in hybrid populations due to intragenic recombination. *Evolution* **37**, 17-29.

Gutz, H., and Leslie, J. F. 1976. Gene conversion: a hitherto overlooked parameter in population genetics. *Genetics* **83**, 861-866.

Harary, F. 1969. *Graph Theory.* Addison-Wesley, Reading, Mass.

Hudson, R. R. 1983. Properties of a neutral-allele model with intragenic recombination. *Theor. Pop. Biol.* **23**, 183-201.

Kaplan, N. L., and Brookfield, J. F. Y. 1983. The effect on homozygosity of selective differences between sites of transposable elements. *Theor. Pop. Biol.* **23**, 273-280.

Karlin, S. 1982. Classifications of selection-migration structures and conditions for a protected polymorphism. *Evol. Biol.* **14**, 61-204.

Kempthorne, O. 1957. *An Introduction to Genetic Statistics.* Wiley, New York.

Kimura, M., and Crow, J. F. 1964. The number of alles that can be maintained in a finite pouplation. *Genetics* **49**, 725-738.

Kingman, J. F. C. 1980. *Mathematics of Genetic Diversity.* Soc. Ind. Appl. Math., Philadelphia.

Lamb, B. C., and Helmi, S. 1982. The extent to which gene conversion can change allele frequencies in populations. *Genet. Res.* **39**, 199-217.

Langley, C. H., Brookfield, J. F. Y., and Kaplan, N. L. 1983. Transposable elements in mendelian populations. I. A theory. *Genetics* **104**, 457-471.

Malécot, G. 1946. La consanguinité dans une population limitée. *Compt. Rend. Acad. Sci.* **222**, 841-843.

Malécot, G. 1948. Les mathématiques de l'hérédité. Masson, Paris. (Extended translation: *The Mathematics of Heredity*. Freeman, San Francisco, 1969).

Malécot, G. 1951. Un traitement stochastique des problèmes linéaires (mutation, linkage, migration) en Génétique de Population. *Ann. Univ. Lyon*, Sci., Sec. A, **14**, 79-117.

Maruyama, T. 1971. An invariant property of a structured population. *Genet. Res.* **18**, 81-84.

Maruyama, T. 1972. Some invariant properties of a geographically structured population: distribution of heterozygotes under irreversible mutation. *Genet. Res.* **20**, 141-149.

Maruyama, T. 1977. *Stochastic Problems in Population Genetics*. Springer, Berlin.

Moody, M. 1981. Polymorphism with selection and genotype-dependent migration. *J. Math. Biol.* **11**, 245-267.

Moran, P. A. P. 1962. *The Statistical Processes of Evolutionary Theory*. Clarendon Press, Oxford.

Nagylaki, T. 1977a. *Selection in One- and Two-Locus Systems*. Springer, Berlin.

Nagylaki, T. 1977b. Decay of genetic variability in geographically structured populations. *Proc. Natl. Acad. Sci. USA* **74**, 2523-2525.

Nagylaki, T. 1978. The geographical structure of populations. In *Studies in Mathematics*. Vol. 16: *Studies in Mathematical Biology*. Part II (S. A. Levin, ed.). Pp. 588-624. The Mathematical Association of America, Washington.

Nagylaki, T. 1980. The strong-migration limit in geographically structured populations. *J. Math. Biol.* **9**, 101-114.

Nagylaki, T. 1982. Geographical invariance in population genetics. *J. Theor. Biol.* **99**, 159-172.

Nagylaki, T. 1983a. Evolution of a large population under gene conversion. *Proc. Natl. Acad. Sci. USA* **80**, No. 19 (in press).

Nagylaki, T. 1983b. Evolution of a finite population under gene conversion. *Proc. Natl. Acad. Sci. USA* **80**, No. 20 (in press).

Nagylaki, T. 1983c. The evolution of multigene families under intrachromosomal gene conversion. Submitted to *Genetics*.

Nagylaki, T. 1983d. The robustness of neutral models of geographical variation. *Theor. Pop. Biol.* **24** (in press).

Nagylaki, T., and Lucier, B. 1980. Numerical analysis of random drift in a cline. *Genetics* **94**, 497-517.

Nagylaki, T., and Moody, M. 1980. Diffusion model for genotype-dependent migration. *Proc. Natl. Acad. Sci. USA* **77**, 4842-4846.

Nagylaki, T., and Petes, T. D. 1982. Intrachromosomal gene conversion and the maintenance of sequence homogeneity among repeated genes. *Genetics* **100**, 315-337.

Ohta, T. 1976. A simple method for treating the evolution of multigene families. *Nature* **263**, 74-76.

Ohta, T. 1982. Allelic and nonallelic homology of a supergene family. *Proc. Natl. Acad. Sci. USA* **79**, 3251-3254.

Ohta, T. 1983a. On the evolution of multigene families. *Theor. Pop. Biol.* **23**, 216-240.

Ohta, T. 1983b. Time until fixation of a mutant belonging to a multigene family. *Genet. Res.* **41**, 47-55.

Ohta, T., and Dover, G. 1983. Population genetics of multigene families, that are dispersed into two or more chromosomes. *Proc. Natl. Acad. Sci. USA* **80**, 4079-4083.

Rusinek, R. 1982. Rate of extinction and limiting distribution for a geographically structured population. *SIAM J. Appl. Math.* **42**, 86-93.

Sawyer, S. 1976. Results for the stepping-stone model for migration in population genetics. *Ann. Prob.* **4**, 699-728.

Sawyer, S. 1977a. Asymptotic properties of the equilibrium probability of identity in a geographically structured population. *Adv. Appl. Prob.* **9**, 268-282.

Sawyer, S. 1977b. Rates of consolidation in a selectively neutral migration model. *Ann. Prob.* **5**, 486-493.

Sawyer, S. 1978. Isotropic random walks in a tree. *Z. Wahr.* **42**, 279-292.

Sawyer, S. 1979. A limit theorem for patch sizes in a selectively neutral migration model. *J. Appl. Prob.* **16**, 482-495.

Sawyer, S., and Felsenstein, J. 1983. Isolation by distance in a hierarchically clustered population. *J. Appl. Prob.* **20**, 1-10.

Slatkin, M. 1981. Fixation probabilities and fixation times in a subdivided population. *Evolution* **35**, 477-488.

Smith, G. P. 1973. Unequal crossing over and the evolution of multigene families. *Cold Spring Harbor Symp. Quant. Biol.* **38**, 507-513.

Tartof, K. 1973. Unequal mitotic sister-chromatid exchange and disproportionate replication as mechanisms regulating ribosomal RNA gene redundancy. *Cold Spring Harbor Symp. Quant. Biol.* **38**, 491-500.

Walsh, J. B. 1983. Role of biased gene conversion in one-locus neutral theory and genome evolution. *Genetics*, in press.

Weinberger, H. F. 1982. Long-time behavior of a class of biological models. *SIAM J. Math. Analysis* **13**, 353-396.

Wright, S. 1969. *Evolution and the Genetics of Populations*. Vol. 2. The University of Chicago Press, Chicago.

Yanagida, E. 1982. Stability of stationary distributions in a space-dependent population growth process. *J. Math. Biol.* **15**, 37-50.

DEPARTMENT OF BIOPHYSICS AND THEORETICAL BIOLOGY
THE UNIVERSITY OF CHICAGO
920 EAST 58TH STREET
CHICAGO, ILLINOIS 60637

Proceedings of Symposia in Applied Mathematics
Volume 30, 1984

EVOLUTION: GAME THEORY AND ECONOMICS

Ethan Akin

Not man alone but all organisms face the economic problem. Limited resources of time, energy and environmental provender must be allocated among the demands of maintenance and reproduction. In recent years biologists have been adapting theories of the human economy to the problems of biological populations.

A population is a subset of a species which is sufficiently isolated that intraspecific interactions, eg. mating, take place amongst members of the subset. We will assume the population is large enough that we can usefully measure the numbers or mass of subpopulations by continuous variables.

In our models the individuals of the population have a choice of actions or strategies from a set I. I will either be a finite set of discrete alternatives or the continuous range of one or more real variables. Each $i \in I$ has an associated payoff A_i which it is in the individual's interest to maximize. Instead of selection based on calculation and subsequent choice we assume that the pattern of strategies in the population is shaped by natural selection. The relatively less successful strategies are gradually eliminated from the population as they are reproduced less frequently. Thus, the payoffs A_i will often be measured in terms of "fitness" or net reproductive rates.

We will concentrate on single populations and ignore inter-population effects like competition and predation. The examples which remain break roughly into two classes according to whether or not there is feedback of the population strategy choices upon the payoff values.

Simple Optimization Models: In these cases the A_i's depend on environmental conditions which may vary with time (eg. weather) or with the total population size (density-dependence) but they do not depend on the frequency of the strategy choices in the population. Finding the optimum is then a classical economic calculation of costs versus benefits.

A widely studied example concerns life history strategies. Suppose the population is divided into age classes with x_α the size of age class α $(= 1,\ldots,n)$. Writing $X(t)$ for the column vector of these age classes at time t, the population dynamic is assumed to be given by the Leslie matrix model described by Dr. Frauenthal:

$$(1) \qquad X(t+1) = LX(t)$$

$$L = \begin{pmatrix} m_1 & \cdots & & m_n \\ p_1 & 0\cdots & & 0 \\ 0 & & & \\ 0 & & \cdots p_{n-1} & 0 \end{pmatrix}$$

where m_α and p_α are, respectively, the average fecundity and survival probability of an individual of age α. Defining $\ell_1 = 1$ and

$\ell_\alpha = p_1 \cdots p_{\alpha-1}$ $(\alpha = 2,\ldots,n)$ the dominant eigenvalue of the non-negative matrix L is the positive root of:

$$\Sigma_\alpha \lambda^{-\alpha} \ell_\alpha m_\alpha = 1. \tag{2}$$

This eigenvalue λ is called the asymptotic growth rate. As discussed by Dr. Frauenthal, the distribution of age classes approaches that of the normalized right eigenvector of L: the stable age distribution. When the population is at the stable age distribution each age class grows geometrically with ratio λ.

Equation (2)--essentially the characteristic equation--defines λ as an implicit function of the entries of L. The strategies are alternate matrices L_i and the payoff is $\lambda(L_i)$. The subpopulation with the largest λ outcompetes the rest by growing--eventually--at a faster rate.

While λ is a monotone function of the m_α's and p_α's, there are assumed to be biological constraints on these variables. For example, a trade off between reproductive effort and maintenance at age α can be represented by a constraint $F(m_\alpha, p_\alpha) \leq C$ where F is monotone in each variable. A detailed analysis due to W.D. Hamilton is described by Roughgarden [5; Chapter 19]. In economic language p_α is increased until the marginal benefit of further increase is equal to the marginal cost of the associated decrease in m_α where value is measured by λ.

Density dependence is introduced by assuming that the m_α's and p_α's are decreasing functions of the population size $x = \Sigma_\alpha x_\alpha$. λ then becomes a decreasing function of x and so, assuming $\mathrm{Lim}_{x\to 0^+} \lambda(x) > 1 > \mathrm{Lim}_{x\to\infty} \lambda(x)$, we get a unique value $\hat{x}$, called the <u>carrying capacity</u>, at which $\lambda(\hat{x}) = 1$. Through the dependence of L upon $x(t)$ system (1) becomes nonlinear and has a unique nonzero equilibrium with population size $\hat{x}$ and stable age distribution for $L(\hat{x})$.

The strategies are now alternate matrix functions L_i where the common independent variable x applied to all the L_i's in the population is the population size summed over all age classes of all strategies. The payoff is $\hat{x}(L_i)$. Suppose that L_{i*} has the largest carrying capacity. When $x < \hat{x}(L_{i*})$ then $\lambda(L_{i*}) > 1$ and the $i*$ subpopulation increases while once x is greater than the remaining $\hat{x}(L_j)$'s, $\lambda(L_j) < 1$ and the remaining subpopulations decrease.

<u>Models with Frequency Dependence</u>: These models exhibit strategic feedback. If p is the distribution of strategies in the population then the payoff A_i is a function of p. In contrast to simple optimization models, strategic feedback imports into the biological context all of the complexities of noncooperative games. This subject is not so much a theory as a folklore of interesting examples illustrating the difficulty of even defining a solution or optimum. Maynard Smith's great contribution was not merely to introduce game theory to biology but to define a useful and general solution idea: the evolutionarily stable state (ESS). A distribution of strategies is evolutionarily stable if it does better against any slightly perturbed distribution than the perturbed distribution itself. Formally, if $A_i(p)$ is the payoff to strategy i, when the population is in state p, then the payoff to a distribution q is its expected value:

$$A_q(p) = \int_I A_i(p)q(di) = \Sigma_I A_i(p)q_i$$

where the latter sum formulation applies when I is finite. So p^* is an ESS if for all p near p^*

(3) $$A_{p^*}(p) \geq A_p(p)$$

with equality only when $p = p^*$. By writing $p = (1-\varepsilon)p^* + \varepsilon q$ for an arbitrary distribution q and $\varepsilon > 0$ we can use linearity in the subscript variable to get the equivalent condition of noninvadability: for all distributions q and all $\varepsilon > 0$ sufficiently small

$$A_{p^*}((1-\varepsilon)p^* + \varepsilon q) \geq A_q((1-\varepsilon)p^* + \varepsilon q),$$

with equality only when $q = p^*$.

While Maynard Smith's definition was static the implied dynamic was made explicit by Taylor and Jonker [20] in the case where I is finite. Let x_i be the size of the i-strategist subpopulation so that $x = \Sigma_I x_i$ is the total population size. Define $p_i = x_i/x$ so that p lies in the simplex $\Delta = \{p \in \mathbb{R}^I: p_i \geq 0 \text{ and } \Sigma p_i = 1\}$. We assume that the payoff $A_i(p)$ is the relative reproductive rate and so:

(4) $$\frac{dx_i}{dt} = x_i A_i(p) \quad \text{and} \quad \frac{dx}{dt} = xA_p(p),$$

where the latter equation is obtained by summation since $x_i = xp_i$. Because $d\,\ell n\, p_i = d\,\ell n\, x_i - d\,\ell n\, x$ we can subtract to get:

(5) $$\frac{dp_i}{dt} = p_i(A_i(p) - A_p(p)).$$

Taylor and Jonker observed that an ESS is always a locally stable equilibrium point for the dynamical system (5). This result was sharpened by Hofbauer, Schuster and Sigmund to the following theorem whose statement requires a bit of game theory terminology. For any distribution $p \in \Delta$, a strategy $i \in I$ is called <u>active</u> for p if it occurs in the population. The set of active strategies is called the <u>support of</u> p, i.e. $\text{supp}(p) = \{i \in I: p_i > 0\}$. The strategies of full support, or interior strategies, consist of the set $\overset{\circ}{\Delta} = \{p \in \Delta: p_i > 0 \text{ for all } i \in I\}$.

<u>1 Theorem</u>: (a) $p^* \in \Delta$ is an equilibrium for (5) if and only if every strategy active for p^* has the same payoff when the population is in state p^*, i.e. $A_i(p^*)$ is the same for all $i \in \text{supp}(p^*)$. This common value is $A_{p^*}(p^*)$. (This condition is due to Bishop and Cannings.)

(b) For any $p^* \in \Delta$ the function of p: $I^{p^*}(p) = -\Sigma\, p_i^*\, \ell n(p_i/p_i^*)$ (summation over the support of p^*) is a smooth, nonnegative function on the open set $\{p \in \Delta: \text{supp}(p) \supset \text{supp}(p^*)\}$. $I^{p^*}(p) = 0$ precisely when $p = p^*$.

(c) $p^* \in \Delta$ is an ESS if and only if it is an equilibrium for (5) and, in addition, I^{p^*} is a local Lyapunov function, i.e. on some neighborhood of p^*

$$\frac{dI^{p^*}}{dt} \leq 0 \tag{6}$$

with equality only at p^*.

<u>Proof</u>: (a) From (4) p^* is an equilibrium if and only if $A_i(p^*) = A_{p^*}(p^*)$ for all $i \in \text{supp}(p^*)$. But if $A_i(p^*) = c$ for all

$i \in \text{supp}(p^*)$ then $C = \Sigma\, p_i^* C = \Sigma\, p_i^* A_i(p^*) = A_{p^*}(p^*)$ because the summations are the same carried out over supp(p*) or all of I.

(b) By concavity of the log function:

$$I^{p^*}(p) \geq -\ell n(\Sigma\, p_i^*(p_i/p_i^*)) \geq -\ell n(1) = 0$$

with equality only when p_i and p_i^* have a common ratio for $i \in \text{supp}(p^*)$ and this support equals that of p, i.e. p* = p.

(c) (6) follows immediately from (3) because at p:

$$\text{(7)} \qquad \frac{dI^{p^*}}{dt} = \Sigma\, p_i^*(A_p(p) - A_i(p)) = A_p(p) - A_{p^*}(p).$$

Again, summation over supp(p*) is the same as summation over I. From this it follows that p* is a locally stable equilibrium. The direct proof that an ESS is an equilibrium is interesting in itself and we will return to it later. QED

The application of both sorts of models raise issues which are currently controversial in evolutionary theory. Some authors object that optimization attributes a degree of perfection to biological adaptation more appropriate to the operation of religious design than to the piecemeal, myopic process of natural selection (see [3]). There may be additional biological constraints beyond those apparent in the structure of the model. The obverse of this same objection is that the degrees of freedom remaining in the model must actually be available as real or potential variability in the population for natural selection to act upon if motion to the optimum is to be achieved. In my view this objection says little more than that speculative predictions require empirical testing before they are taken too seriously. Other more technical problems are suggested by the models themselves. For example, in the density dependent age-structure case I glossed over the question of local stability of the equilibrium and the possibility of nonequilibrium attractors, eg. limit cycles. When these occur competition between competing subpopulations using different strategies can no longer necessarily be described by maximizing some payoff function.

The other major issue concerns the level at which selection acts. Instead of the harmonious progression of adaptive traits selected "for the good of the species" we hear recently the more dissonant theme of conflict of interests between individuals in the population. The divergence of Darwinian natural selection from presumed species benefit is revealed most sharply in the problem, first raised by Haldane [2], of altruistic traits.

We can illustrate the problem of altruism using the frequency dependent model of equations (4) and (5). I consists of two strategies: $\{1,2\}$ with 1 the altruistic, or cooperative, trait and 2 the cheating, or noncooperative, trait. So p_1 is the fraction of altruists in the population, and since the vector $p = (p_1,p_2) = (p_1,1-p_1)$ we can regard the A_i's as functions of p_1 $(i = 1,2)$. So (4) and (5) become

$$\frac{dx}{dt} = xA_p(p) \quad \text{with } A_p(p) = p_1A_1(p_1) + (1-p_1)A_2(p_1)$$

$$\frac{dp_1}{dt} = p_1(A_1(p)-A_p(p)) = p_1(1-p_1)(A_1(p_1)-A_2(p_1)).$$

The phenomenon of altruism arises when:

(8a) $$A_i'(p_1) \equiv \frac{dA_i}{dp_1} > 0 \quad \text{for all values of } p_1 \quad (i = 1,2).$$

(8b) $$A_2(p_1) > A_1(p_1) \quad \text{for all values of } p_1.$$

Condition (8a) says that every increase in the frequency of altruists increases the fitness of both altruists and nonaltruists. Condition (8b) says that the nonaltruists always do better, or in game theory parlance, strategy 2 dominates strategy 1. See Figure 1.

From the dynamic viewpoint domination, (8b), is all that matters. Pure altruism, fixation with $p_1 = 1$, is an unstable equilibrium. Any perturbation, any introduction of strategy 2, poisons the population which then tends to fixation with $p_1 = 0$ because $A_1 - A_2 < 0$. But (8a) makes this result appear paradoxical. Each subpopulation does worse and worse as the frequency of altruists decreases. Furthermore, the entire population also does worse if we assume, in addition:

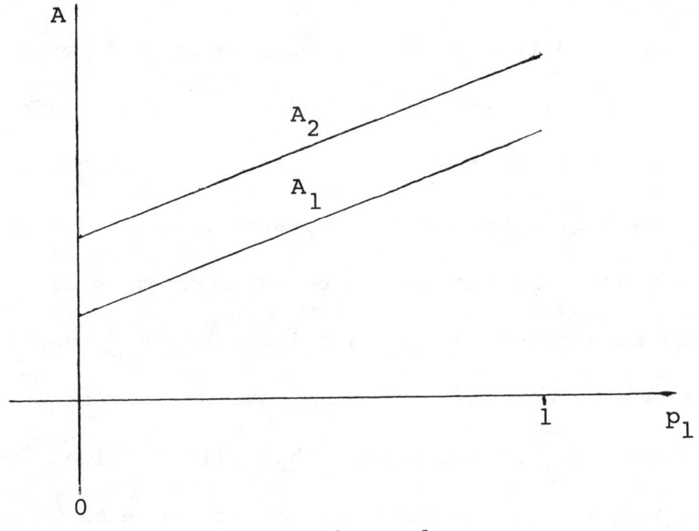

Figure 1

$$\frac{dA_p(p)}{dp_1} > 0 \qquad \text{for all values of } p_1, \text{ i.e.}$$

(9)

$$p_1 A_1'(p_1) + (1-p_1)A_2'(p_1) > A_2(p_1) - A_1(p_1).$$

In Figure 1 this says that the average slope exceeds the difference between the two graphs.

While this model is simple it can be used to clarify an example where even "individual benefit" leads to the same sort of pitfalls as "group benefit".

The classic example of presumed altruism is the alarm call: upon spotting a predator I scream and run. The effort of the scream and the risk of attracting the predator's attention are regarded as (possibly small) costs to me, while nearby conspecifics receive the benefit of the warning. Trivers [15] suggests that the call may be to my individual advantage because the predator who successfully hunts one of my species, even if not me or my relatives, may be more likely to concentrate on my kind, putting me at greater risk later (the "search image" hypothesis). This would appear to solve the alarm call problem by reference to its individual benefit. Trivers himself does not commit this beguiling error as he refers to the need for spatial variation, of which more below. However, some other authors, citing his discussion, do fall into the trap. The mistake is revealed by considering competition between strategies rather than individuals. Even if correct, the above argument does not eliminate domination by the noncooperative strategy. It would mean that silent flight is a "spiteful" strategy which hurts me but hurts others more as we share the search image cost while I save the alarm call risk cost. The level of search image cost depends just on the frequency p_1 but the risk cost savings is the difference $A_2(p_1) - A_1(p_1)$ which is positive. So a supplementary explanation is still required. It is Dawkins [14] who especially emphasizes the usefulness of this focus upon strategies rather than individuals.

The honor of raising this issue to modern consciousness belongs to Wynne-Edwards. In [12] he argued that a number of altruistic traits are maintained by group selection, i.e. those populations which are poisoned by an increasing percentage of noncooperators are outcompeted and eventually replaced by populations which have remained purely altruistic. Williams [13] replied that group selection can successfully oppose individual selection only under very unusual circumstances. The argument is based on relative rates. Darwinian selection proceeds by differential births and deaths and so its time scale is the generation. Group selection operates by differential extinctions and colonizations of entire populations and so has a time scale an order of magnitude slower. Furthermore, isolation between the competing populations is required. Virtually any migration between populations will cause the infection of noncooperation to spread laterally between populations before prophylactic extinction can eliminate it.

The current view is that the stability of an altruistic trait requires some device whereby its possessor receives proportionally more of the benefits of the altruistic acts than does the average member of the population. This can be accomplished by donor discrimination in favor of other altruists (Trivers' reciprocal altruism) or in favor of relatives (Hamilton's kin selection). In the latter case kin of altruists are more likely than average to carry the altruistic trait while kinship may be easier to recognize than possession of the trait itself.

Stability can also result if the trait shows variation of occurrence in a spatially extended population. This mechanism is interesting because it does not require behavioral discrimination by the donors. D.S. Wilson [17] shows how this can happen as a consequence of a statistical result which we may call the Redhead Theorem: redheads in a population tend to see more redheads than the average person.

Suppose a population of size x is divided into subpopulations with x_α the size of α subpopulation, $x = \Sigma\, x_\alpha$. Define $q_\alpha = x_\alpha/x$. Suppose there are $x_{1\alpha}$ altruists in population α so that $p_{1\alpha} = x_{1\alpha}/x_\alpha$ is the frequency there. The mean frequency p_1 is given by:

$$p_1 = \Sigma\, q_\alpha p_{1\alpha} = \frac{\Sigma\, x_\alpha (x_{1\alpha}/x_\alpha)}{\Sigma\, x_\alpha} = \frac{\Sigma\, x_{1\alpha}}{\Sigma\, x_\alpha}\,. \tag{10}$$

As an individual is in population α with probability q_α, equation (10) says that p_1 is the frequency of altruism observed by a random member of the population. Now consider $p_1^{(1)}$ the frequency of altruism as observed by a random altruist. Because $x_{1\alpha} = x_\alpha p_{1\alpha}$ we have

$$p_1^{(1)} = \frac{\Sigma\, x_{1\alpha}\, p_{1\alpha}}{\Sigma\, x_{1\alpha}} = \frac{\Sigma\, x_\alpha\, p_{1\alpha}^2}{\Sigma\, x_\alpha\, p_{1\alpha}} = \frac{\Sigma\, q_\alpha\, p_{1\alpha}^2}{\Sigma\, q_\alpha\, p_{1\alpha}} = p_1 + (\sigma^2/p_1) \tag{11}$$

where σ^2 is the variance of $p_{1\alpha}$ computed via the distribution q_α. To see this another way, note that, by Bayes' Theorem, an altruist occurs in population α with probability

$$q_\alpha^{(1)} = q_\alpha p_{1\alpha}/p_1 = x_{1\alpha}/\Sigma\, x_{1\alpha}$$

and (11) says that $p_1^{(1)} = \Sigma\, q_\alpha^{(1)} p_{1\alpha}$. Similarly, the frequency of altruism as observed by a nonaltruist is computed via $q_\alpha^{(2)} = q_\alpha(1-p_{1\alpha})/(1-p_1)$:

$$(12) \qquad p_1^{(2)} = \Sigma\, q_\alpha^{(2)} p_{1\alpha} = p_1 - (\sigma^2/1-p_1)$$

If the effect of the altruistic acts are restricted to subpopulations then the fitness of strategy i in population α is $A_i(p_{1\alpha})$. So, at least in the linear case illustrated by Figure 1, the average fitness of an altruist is $A_1(p_1^{(1)})$ and of a nonaltruist is $A_2(p_1^{(2)})$. As $p_1^{(2)} < p_1^{(1)}$ with difference $\sigma^2/p_1(1-p_1)$ it is easy to see graphically that $A_1(p_1^{(1)})$ can be larger than $A_2(p_1^{(2)})$ despite domination.

So far this analysis says that the predominantly altruistic subpopulations may do sufficiently well that the frequency of the altruistic trait increases. As Hamilton points out in a related article [16] the requirements for stability are shifted in an interesting way. Competition between isolated groups is not required. In fact, the groups can break up and reform after a dispersal stage. What is required is that the variance in the trait which is diminished by selection be regenerated in some fashion. Wilson shows by example that the variance generated by binomial sampling alone in the dispersal stage may suffice.

For the remainder of the discussion I want to focus on the special case of (5) where the payoffs $A_i(p)$ are linear. This means that there is a constant matrix a_{ij} representing the payoff to the i player when he meets a j-opponent. $A_i(p) = a_{ip} \equiv \Sigma p_j a_{ij}$ and $A_p(p) = a_{pp} \equiv \Sigma p_i p_j a_{ij}$. In this notation (5) becomes:

$$\frac{dp_i}{dt} = p_i(a_{ip} - a_{pp}). \tag{13}$$

Thus, every $I \times I$ matrix a_{ij} yields a vectorfield X^a on Δ defined by the right side of the above equation.

For example, the problem of altruism in the linear version considers a 2×2 matrix a_{ij}. Conditions (8b) and (9) reduce to:

$$a_{21} > a_{11} > (a_{21}+a_{12})/2 > a_{22} > a_{12}, \tag{14}$$

and these imply (8a). To check the equivalence recall than an inequality linear in p_1 is true for all values of p_1 if and only if it is true for the extreme values $p_1 = 0,1$. This two-person game is the celebrated Prisoner's Dilemma.

The dynamical system (13) arises in a number of different contexts, eg. origin of life problems [22] and population genetics. The discrete time version with an antisymmetric matrix is the model of gene conversion discussed by Professor Nagylaki.

The special case where the matrix is symmetric, i.e.

$$a_{ij} = a_{ji} \tag{15}$$

is the classical genetic model of natural selection for a large, sexual population. In that context I is the set of alternative genes (alleles) at a position on the chromosome (locus). a_{ij} is the fitness of an ij zygote and so is symmetric by definition. From the game theory perspective symmetry means that the two players always get the same payoff. The game appears to be totally cooperative and the problem is to find where both players do best. The result which verifies this intuition is known as Fisher's Fundamental Theorem of Natural Selection. Writing the mean payoff a_{pp} as $\bar{a}$ it is easy to check that

$$\text{(16)} \qquad \frac{d\bar{a}}{dt} = 2\Sigma\, p_i(a_{ip} - a_{pp})^2 \geq 0 \qquad \text{(given (15))}$$

with equality only at an equilibrium.

Furthermore, the Maximum Principle of Kimura says that the direction of the vectorfield, in the symmetric case, is that of greatest increase of $\bar{a}$. This suggests that X^a is, up to a positive multiple, the gradient of $\bar{a}$. However the ordinary gradient of $\bar{a}$ is not parallel to X^a. This corresponds to a peculiarity in Kimura's proof--which biologists have remarked upon--concerning the meaning of a unit direction. The solution of this puzzle is due to Shahshahani and Conley. To get the gradient of a function you begin with its differential and take the dual vectorfield using a Riemannian metric. So the concept of gradient depends upon the choice of Riemannian metric. The Shahshahani metric is a Riemanian metric on $\mathring{\Delta}$ different from the usual Euclidean metric.

Consider a vectorfield X on $(\mathbb{R}^I)_+ = \{x \in \mathbb{R}^I : x_i > 0$ for all $i\}$. The associated differential equation is:

$$\frac{dx_i}{dt} = X_i = x_i \xi_i$$

so that X_i is the absolute rate of increase of x_i and $\xi_i \equiv X_i/x_i$ is the relative rate of increase. The Euclidean notion of length is given by

$$\|X\|_0^2 = \Sigma X_i^2 = \Sigma x_i^2 \xi_i^2$$

which is independent of the point x at which the vector is based. The Shahshahani notion of length is

$$\|X\|_x^2 = \Sigma X_i^2/x_i = \Sigma x_i \xi_i^2.$$

More generally if X and Y are vectors based at x with components $X_i = x_i \xi_i$ and $Y_i = x_i \eta_i$ then the Shahshahani inner product is defined by:

$$(X,Y)_x = \Sigma X_i Y_i/x_i = \Sigma x_i \xi_i \eta_i. \tag{17}$$

The Shahshahani metric has proved to be natural for a number of problems. For example, let U be a real valued function on $(\mathbb{R}^I)_+$. The vectorfield X is the Euclidean gradient of U if its components, the absolute rates, are the partials of U, i.e. $X_i = \frac{\partial U}{\partial x_i}$. It is the Shahshahani gradient of U if the relative rates are the partials, i.e. $\xi_i = \frac{\partial U}{\partial x_i}$. So in models which begin by describing relative rates of increase, the Shahshahani geometry is likely to be the natural one.

If we restrict to the open simplex $\mathring{\Delta} = \Delta \cap (\mathbb{R}^I)_+$ then the unrestricted gradient must be projected onto the tangent space of Δ which is $\mathbb{R}_0^I = \{X: \Sigma X_i = 0\}$. So the Shahshahani gradient on $\mathring{\Delta}$ of U at p is given by:

(18) $\qquad (\bar{\nabla}_p U)_i = p_i(\frac{\partial U}{\partial x_i} - \overline{\frac{\partial U}{\partial x}}) \qquad$ (evaluated at $x = p$)

where $\overline{\frac{\partial U}{\partial x}} \equiv \Sigma\, p_i \frac{\partial U}{\partial x_i}$.

Returning to equation (13) with a_{ij} symmetric it is easy to check from (18) that the vectorfield X^a is the Shahshahani gradient of $\bar{a}/2$. Thus, selection of gene frequencies acts to optimize mean fitness $\bar{a}$. Recall that $\bar{a}$ is the population growth rate, i.e. the second equation of (4) is:

(19) $\qquad \frac{dx}{dt} = x\bar{a}.$

We can introduce density dependence, as we did in the age-structured model, by assuming the a_{ij}'s are functions of x as follows

$$\frac{da_{ij}}{dx} < 0 \qquad a_{ij} = a_{ji} \text{ for all } x$$

$$\text{Lim}_{x\to 0+}\, a_{ij} > 0 > \text{Lim}_{x\to\infty}\, a_{ij}.$$

So for each $p \in \Delta$ we can define the associated carrying capacity $\hat{x} = \hat{x}(p)$ by the equation

(20) $\qquad 0 = \bar{a}(\hat{x}) \qquad (= \Sigma\, p_i p_j a_{ij}(\hat{x})).$

Notice that $\bar{a} > 0$ (resp. < 0) if and only if $x < \hat{x}(p)$ (resp. $x > \hat{x}(p)$).

If we take the gradient of the equation (20) we get

$$\bar{\nabla}_p \bar{a} = \frac{\partial \bar{a}}{\partial x} \bar{\nabla}_p \hat{x} = 0$$

where $\bar{\nabla}_p \bar{a}$ is the gradient with respect to the p variables with

x then fixed at $\hat{x}(p)$ and $\frac{\partial \bar{a}}{\partial x} = \Sigma\, p_i p_j \frac{da_{ij}}{dx} < 0$. This means that when $x = \hat{x}(p)$ the gradient of $\hat{x}$ points in the same direction as the partial gradient of $\bar{a}$ $(= 2X^a)$. This implies the following which can also be computed directly from (20):

$$\frac{d\hat{x}}{dt} = \frac{2}{\left|\frac{\partial \bar{a}}{\partial x}\right|} \|\bar{\nabla}_p(\bar{a}/2)\|_p^2 = \frac{2}{\left|\frac{\partial \bar{a}}{\partial x}\right|} \Sigma\, p_i (a_{ip} - a_{pp})^2 \qquad (21)$$

$$(\text{when } x = \hat{x}(p)).$$

Note that this is nonnegative, vanishing only at equilibrium. Following Ginzburg [6] one can then show that the system tends to equilibrium and that (p^*, x^*) is a locally stable equilibrium if and only if $x^* = \hat{x}(p^*)$ and this value is a local maximum for the function $\hat{x}$. As before density dependence replaces growth rate by carrying capacity as the objective of optimization. Notice that if we replace the dynamic adjustment (19) by instantaneous size adjustment: $x = \hat{x}(p)$, then $\hat{x}(p)$ is a Lyapunov function for the new system by (21).

Let us return to the density independent case. A vectorfield X on Δ is just a function $X\colon \Delta \to \mathbb{R}_0^I$. By differentiating at $p \in \mathring{\Delta}$ we define a bilinear form called the <u>Hessian</u> of X at p:

$$H_p(X)(Y_1, Y_2) = (D_pX(Y_1), Y_2)_p \quad (Y_1, Y_2 \in \mathbb{R}_0^I). \qquad (22)$$

The Hessian is a useful tool for two reasons. The vectorfield X is the Shahshahani gradient of some function if and only if the Hessian is a symmetric form at every point $p \in \mathring{\Delta}$. Secondly, if p is an equilibrium, i.e. $X(p) = 0$, then the derivative at p, $D_pX\colon \mathbb{R}_0^I \to \mathbb{R}_0^I$, is the linearization of X at p whose eigenvalues determine local stability. The Hessian is the bilinear form dual with respect to the Shahshahani metric. An application of the first property is the following (cf. Akin [23]):

2 <u>Theorem</u>. Let X be any vectorfield on $\mathring{\Delta}$ which is not a gradient. In the family of vectorfields $\{X^a + X\}$, where a_{ij} varies over all symmetric matrices, Hopf bifurcation takes place and so there occurs examples with nonconstant periodic solutions.

<u>Sketch of proof</u>: Because X is not a gradient there exists $p^* \in \mathring{\Delta}$ such that $H_{p^*}(X)$ is not symmetric. The family of symmetric matrices is just large enough that a unique a_{ij} is determined by (1) the restriction that p^* be an equilibrium for $X^a + X$ and (2) the choice of symmetric part for the form $H_{p^*}(X^a + X)$. Furthermore, the choice in (2) is arbitrary. So one can define a one parameter family a^ϵ such that p^* remains at equilibrium and one conjugate pair of eigenvalues crosses the imaginary axis as ϵ varies. This leads to Hopf bifurcation and hence to cycles. QED

In population genetics we let X model the effect of recombination between two loci and obtain the unexpected occurrence of cycles in the two-locus-two-allele model. In game dynamics we let $X = X^b$ with b_{ij} antisymmetric and get cycling examples for (13). However, the nature of the Hopf bifurcation requires more investigation in each example. It can happen that the Hopf bifurcation is degenerate so that the cycling examples are not robust. For example, Zeeman shows this is true for the three strategy examples of (13) [30]. In the four strategy case he constructs nondegenerate examples leading to robust limit cycles. Such limit cycles can also occur in the two-locus-two-allele model (see [24]).

If $p^* \in \mathring{\Delta}$ is our equilibrium for X^a then it is easy to check that

$$(23) \qquad H_{p^*}(X^a)(Y_1, Y_2) = \Sigma\, a_{ij} Y_{1j} Y_{2i} \qquad (Y_1, Y_2 \in \mathbb{R}^I_0)$$

which will usually not be symmetric if the matrix a_{ij} is not. Notice that the form does not depend on the point p^*. However, the eigenvalues of the linearization do depend on the point because to get the matrix of the linearization from the form one uses a $(\, , \,)_{p^*}$ orthonormal basis for $\mathbb{R}^I_0$. In practice we care less about the precise eigenvalues than whether they all have negative real parts, which is the condition for local stability. A classic result of Lyapunov says that a sufficient, but certainly not necessary, condition for stability is that the symmetrization be negative definite:

$$(24) \qquad H_{p^*}(Y,Y) = \Sigma\, a_{ij} Y_i Y_j < 0 \qquad \text{for all } Y \in \mathbb{R}^I_0,\ Y \neq 0.$$

For an interior equilibrium this is exactly the condition that p^* be an ESS.

3 <u>Theorem</u>. Let $p^* \in \mathring{\Delta}$. p^* is an ESS for a_{ij} if and only if p^* is an equilibrium for X^a and the Hessian is negative definite at p^*, i.e. (24) holds.

<u>Proof</u>: Fix $Y \in \mathbb{R}^I_0$ and substitute $p = p^* + \epsilon Y$ with $\epsilon > 0$ sufficiently small into inequality (3). Expanding out and canceling we get $0 > a_{Yp^*} + \epsilon a_{YY}$. In particular, $0 \geq a_{Yp^*}$ for all $Y \in \mathbb{R}^I_0$.

Replacing Y by $-Y$ we get $0 = a_{Yp*}$ and this implies a_{ip*} is independent of i, i.e. $p*$ is an equilibrium. Cancelling ϵ we have $0 > a_{YY} = H_{p*}(Y,Y)$. As this reasoning is reversible we have the equivalence. QED

The dynamic stability of an ESS has already been remarked in Theorem 1. For an interior ESS inequality (6) holds on the entire interior $\mathring{\Delta}$ and so $p*$ attracts all interior initial values. To see this note that equation (7) can be rewritten in this case:

$$\frac{dI^{p*}}{dt} = a_{YY} \qquad \text{with} \qquad Y = p - p* \quad (p \in \mathring{\Delta})$$

because $a_{Yp*} = 0$ by the equilibrium condition. In [30] Zeeman provides an example of a locally stable equilibrium which is not globally stable and so is not an ESS.

As another example of the special role of ESS note first that replacing a_{ij} by $a_{ij} - \epsilon_j$ has no effect on equation (13). In contrast, replacing a_{ij} by $a^{\epsilon}_{ij} \equiv a_{ij} - \epsilon_i$ does change the dynamics. The biological interpretation is that different strategies may have different energetic costs independent of strategic considerations of how i does against j. The perturbation from a to a^{ϵ} is a change in these entry costs. If $p* \in \mathring{\Delta}$ is a nondegenerate equilibrium for X^a, eg. an interior ESS, and the perturbing ϵ_i's are small then $X^{a^{\epsilon}}$ will have a unique equilibrium p^*_{ϵ} near $p*$. The Hessians $H_{p*}(X^a)$ and $H_{p^*_{\epsilon}}(X^{a^{\epsilon}})$ are equal because $\Sigma\, \epsilon_i Y_{1j} Y_{2i} = 0$ as $Y_1 \in \mathbb{R}^I_0$. So if $p*$ is an ESS for a_{ij}, p^*_{ϵ} will be an ESS for a^{ϵ}_{ij} as long as the perturbed equilibrium remains in $\mathring{\Delta}$. While I have not examined the matter I presume that sufficient variation of this sort may destroy local stability for an equilibrium which is not ESS. In fact, I conjecture that for judiciously chosen a_{ij}, Hopf bifurcation can occur in the family $\{X^{a^\epsilon}\}$.

A final point concerns the interpretation of mixed states, eg. $p \in \mathring{\Delta}$. The Taylor-Jonker dynamic (13) is appropriate to polymorphism: individuals play pure strategies and p_i is the fraction of i-strategists in the population. In [26] Hines describes the mathematics when individuals themselves play mixed strategies. Then each individual is associated with a point in Δ and the population is described by a measure on Δ. To see the difference suppose the population is monomorphic for some mixed strategy $p \in \Delta$. Because we are ignoring mutation and every individual breeds true there is no variation upon which selection can act. So the population remains at p regardless of strategic considerations. In all of these models monomorphism is an equilibrium, though it is often unstable. Mathematically, the mixed strategy dynamic leads to interesting extensions of the Shahshahani geometry to spaces of measures. Again ESS plays a special role (see [25] and [29]).

In sum, economic and strategic concepts, especially the idea of ESS, are playing a growing role in biological thinking with many opportunities for mutualism between biological and mathematical intuitions. Even the restricted area of the Taylor-Jonker dynamics has provided an upwelling rich in food for mathematical thought.

BIBLIOGRAPHY

This list is intended as a guide to further reading. It consists mostly of books and surveys from the Quarterly Review of Biology. These tend to have broad bibliographies for those who wish to concentrate on a topic. I have tried to pick out works which I particularly enjoyed. In particular, I recommend the entire series of Princeton Monographs on Population Biology of which several titles appear below.

BACKGROUND: EVOLUTION AND THEORETICAL POPULATION BIOLOGY

[1] J. Maynard Smith: The Theory of Evolution, Penguin Books Inc. (1958).

[2] J.B.S. Haldane: The Causes of Evolution, Cornell U. Press (1966) (a reprint of the 1932 edition).

[3] S.J. Gould: The Panda's Thumb, W.W. Norton and Co., Inc. (1982).

[4] R.H. MacArthur: Geographical Ecology, Harper and Row Publishers (1972).

[5] J. Roughgarden: Theory of Population Genetics and Evolutionary Ecology: An Introduction, MacMillan Publishing Co., Inc. (1979).

[6] L.R. Ginzburg: Theory of Natural Selection and Population Growth, Benjamin/Cummings Publishing Co., Inc. (1983).

[7] R.M. May: Stability and Complexity in Model Ecosystems, Princeton Monographs in Pop. Biol. No. 6, Princeton U. Press (1973).

OPTIMIZATION IN BIOLOGY

[8] S. Stearns: "Life History Tactics: A Review of the Ideas" (1976) Quart. Rev. Biol. 51: 3-50.

[9] R. Levins: Evolution in Changing Environments, Princeton Monographs in Pop. Biol. No. 2, Princeton U. Press (1968).

[10] S.D. Fretwell: Populations in a Seasonal Environment, Princeton Monographs in Pop. Biol. No. 5, Princeton U. Press (1972).

[11] S.A. Levin: "On the Evolution of Ecological Parameters" in Ecological Genetics: The Interface, edited by P.F. Brussard, Springer-Verlag Inc. (1978).

ALTRUISM AND BIOLOGICAL GAMES

[12] V.C. Wynne-Edwards: Animal Dispersion in Relation to Social Behavior, Oliver and Boyd (1962).

[13] G.C. Williams: Adaptation and Natural Selection, Princeton U. Press (1966).

[14] R. Dawkins: The Extended Phenotype, W.H. Freeman and Co. (1982).

[15] R.L. Trivers: "The Evolution of Reciprocal Altruism" (1971) Quart. Rev. of Biol. 46: 35-57.

[16] W.D. Hamilton: "Innate Social Aptitudes of Man: An Approach from Evolutionary Genetics" in Biosocial Anthropology edited by R. Fox, Malaby Press (1975.

[17] D.S. Wilson: The Natural Selection of Populations and Communities, Benjamin/Cummings Publishing Co., Inc. (1980).

[18] J.R. Krebs and N.B. Davies: An Introduction to Behavorial Ecology, Sinauer Associates (1981).

[19] J. Maynard Smith: Evolution and the Theory of Games, Cambridge U. Press (1982).

MATHEMATICS OF EVOLUTIONARY GAMES

[20] P.D. Taylor and L.B. Jonker: "Evolutionarily Stable Strategies and Game Dynamics" (1978) Math. Biosci. 40: 145-156.

[21] P. Schuster, K. Sigmund, J. Hofbauer and R. Wolff: "Self-Regulation in Animal Societies" (I and II) (1981) Biol. Cybern. 40: 1-8 and 9-15.

[22] M. Eigen and P. Schuster: The Hypercycle: A Principle of Natural Self-Organization, Springer-Verlag Inc. (1979).

[23] E. Akin: The Geometry of Population Genetics, Lect.

Notes in Biomath. No. 31, Springer-Verlag (1979).

[24] E. Akin: Hopf Bifurcation in the Two-Locus Genetic Model, to appear Memoirs AMS No. 284 (1983).

[25] E. Akin: "Exponential Families and Game Dynamics" (1982) Can. J. Math. 34: 374-405.

[26] W.G.S. Hines: "Strategy Stability in Complex Populations" (1980) J. Appl. Prob. 17: 600-610.

[27] W.G.S. Hines: "Three Characterizations of Population Strategy Stability" (1980) J. Appl. Prob. 17: 333-340.

[28] W.G.S. Hines: "Mutations, Perturbations and Evolutionarily Stable Strategies" (1982) J. Appl. Prob. 19: 204-209.

[29] E.C. Zeeman: "Population Dynamics from Game Theory", Proc. Int. Conf. Global Theory of Dynamical Systems, Northwestern, Lecture Notes in Math. No. 819, Springer-Verlag (1979).

[30] E.C. Zeeman: "Dynamics of the Evolution of Animal Conflicts" (1981) J. Theoret. Biol. 89: 249-270.

Mathematics Department, The City College of New York,
New York City, N.Y. 10031.

Proceedings of Symposia in Applied Mathematics
Volume 30, 1984

OPTIMAL CONTROL AND PRINCIPLES IN POPULATION MANAGEMENT

WAYNE M. GETZ

ABSTRACT. A wide variety of problems dealing with the growth, management and exploitation of natural resources have been formulated in an optimal control setting. Pontryagin's Maximum Principle [1-4] provides a set of necessary conditions that can be used to construct extremal (candidate) solutions to optimal control problems. Often these solutions embody a qualitative principle. In this paper, problems are presented that demonstrate a particular principle in the growth and management of biological populations. Applications include the exploitation of both lumped and age-structured populations, questions relating to investment, the management of fisheries and pest populations, and the optimal allocation of resources to growth and reproduction. Problems include free, fixed and infinite time mathematical formulations with various types of boundary conditions as well as systems with so-called state variable jump-discontinuities.

1. BACKGROUND. Ordinary differential equations were first used in the early to mid-19th Century to model the growth of a population (Gompertz [5], Verhulst [6]). It is only in the last sixty years, however, that systems of differential equations were used to model population interactions such as predation (Volterra [7], Lotka [8]) and competition (Gause [9]). These models fall within the general autonomous ordinary differential equation framework:

$$\frac{dx}{dt} = f(x) \qquad x(0) = x_0 \tag{1}$$

where $x \in R^n$ and f is a vector of functions $f_i\colon R^n \to R$, $i = 1, \ldots, n$ that possess continuous partial derivatives

$$\frac{\partial f_i}{\partial x_j}, \; i,j = 1, \ldots, n.$$

The elements x_i of x, $i = 1, \ldots, n$, usually denote the density (biomass, numbers) of the i-th species, although in ecosystem models, some of the variables x_i may represent the concentration of nutrients and chemical pollutants. The function, $f_i(x)$, $i = 1, \ldots, n$, characterizes the rate of change of the i-th population x_i (species or chemical) with respect to time.

Until the early seventies, analysis of population models primarily

1980 Mathematics Subject Classification: 92A15, 49B10.

0160-7634/84 $1.00 + $.25 per page

focused on the qualitative properties of solutions to systems where f is explicitly defined or where the behavior of f, usually specified by the sign of its first derivatives, is defined in regions of the positive orthant $x_i \geqslant 0$, $i = 1, \ldots, n$. More recently, emphasis shifted to include the analysis of population and resource management problems in the following optimal control theory framework.

PROBLEM 1. Minimize the objective functional

$$J(u(\cdot)) = \int_0^T L(x,u,t)\, dt + F(x(T)) \tag{2}$$

over all piecewise continuous vector-valued functions $u(\cdot): [0,T] \to U$, where $U \subset R^m$ is a prescribed constraint set, subject to the system constraint equations

$$\frac{dx}{dt} = f(x,u) \tag{3}$$

and boundary conditions

$$x(0) \in \theta_0 \quad \text{and} \quad x(T) \in \theta_1 . \tag{4}$$

The functions $L: R^{m+n+1} \to R$, $F: R^n \to R$ and $f: R^{n+m} \to R^n$ are usually (although not necessarily) assumed to have continuous first derivatives while the boundary sets θ_i, $i = 0,1$, are usually assumed to be nonempty, smooth $(n-q_i)$-dimensional manifolds $(q_i \leqslant n)$ that possess tangent spaces $d_x(\theta_i)$ at every point $x \in \theta_i$. □

The primary tool for analyzing the property of solutions to a particular form of Problem 1 is a maximum principle [1-4,10,11] often referred to as Pontryagin's Maximum Principle. This principle provides a set of necessary conditions that can be used to construct so-called extremal or candidate solutions to the problem on hand. These conditions are stated in terms of the Hamiltonian function

$$H(x,u,\lambda,\lambda_0,t) = \lambda_0 L(x,u,t) + \sum_{i=1}^{n} \lambda_i f_i(x,u), \tag{5}$$

where $\lambda = (\lambda_1,\ldots,\lambda_n)^T$ is termed the costate vector.

MAXIMUM PRINCIPLE. A necessary condition for a control $u^*(\cdot)$ to be a solution to Problem 1 is that there exists a non-trivial vector function $\lambda(\cdot):[0,T] \to R^n$ and constant $\lambda_0 \leqslant 0$ such that an $[0,T]$

$$\text{(i)} \quad \frac{d\lambda_i}{dt} = -\frac{\partial H}{\partial x_i}(x^*(t), u^*(t), \lambda(t), \lambda_0, t) \qquad i = 1, \ldots, n \tag{6}$$

$$\text{(ii)} \quad \max_{u \in U} H(x^*(t),u(t),\lambda(t),\lambda_0,t) = H(x^*(t),u^*(t),\lambda(t),\lambda_0,t), \tag{7}$$

where $x^*(t)$ is a corresponding state vector solution to equation (3) satisfying boundary conditions (4). Further, the costate vector at time $t = 0$, i.e. $\lambda(0)$, is normal to the tangent space $d_{x(o)}(\theta_o)$ while $\lambda(T)$ is the sum of the gradient vector $\lambda_o \frac{\partial F}{\partial x}$ and a vector normal to the tangent space $d_{x(T)}(\theta_1)$. □

For convenience define the so-called switching functions

$$\sigma_i(t) = \frac{\partial H}{\partial u_i}(x^*(t),u^*(t),\lambda(t),\lambda_o,t) \qquad i = 1, \ldots, m \tag{8}$$

If the control constraint set $U = \{u \in R^m | \ a_i \leqslant u_i \leqslant b_i, \ i = 1, \ldots, m\}$ then, using (8), condition (7), for $i = 1, \ldots, n$, becomes

$$u_i^* (t) = a_i \ (b_i) \text{ only if } \sigma_i(t) < (>) \ 0 \tag{9}$$

so that on $(t_1, t_2) \subset [0,T]$

$$u_i^*(t) \in (a_i,b_i) \text{ only if } \sigma_i(t) = 0. \tag{10}$$

Extremal controls that switch between the values expressed in (9) are referred to as *bang-bang* controls.

If the boundary set θ_1 is specified by $q_1 \leqslant n$ equations

$$\psi_j(x) = 0 \qquad j = 1, \ldots, q_1, \tag{11}$$

where $\psi: R^n \to R^{q1}$ is differentiable and the gradient vectors $\frac{\partial \psi_j}{\partial x}$, $j = 1, \ldots q_1$, are linearly independent, then the transversality condition on $\lambda(T)$ can be expressed as

$$\lambda^T (T) = \nu^T \frac{\partial \psi}{\partial x} + \lambda_o \frac{\partial F}{\partial x}^T \tag{12}$$

for some arbitrary vector $\nu \in R^q$. Thus if $x(T)$ is completely specified $(q = n)$ then $\lambda(T)$ is arbitrary while if $x(T)$ is unspecified then $\lambda(T) = - \frac{\partial F}{\partial x}^{\dagger}$. Similarly if $x(0)$ is specified, then $\lambda(0)$ is arbitrary.

If the matrix H_{uu}, with elements $\frac{\partial^2 H}{\partial u_i \partial u_j}$, is singular on $(t_1,t_2) \subset [0,T]$ when evaluated along the optimal solution then the case of *singular control* arises. Of particular interest in population biology is the case where H is linear in u, i.e. $H_{uu} \equiv 0$, and some of the switching functions $\sigma_i(t)$, defined in (8), are zero. In this case conditions (9) and (10) are not directly helpful in the construction of candidate solutions (extremal controls) to Problem 1. Necessary conditions exist involving the time derivatives of $\sigma_i(t)$ [11,12], however, and these can be used to construct extremal controls.

In some problems the value of T is not fixed and the objective functional J has a minimum with respect to T and $u(\cdot)$. Assuming that $\theta_1 = R^n$, i.e.

---

†Under certain controllability assumptions which are satisfied whenever $x(T)$ is unspecified, the control problem is said to be normal and we can set $\lambda_o = - 1$.

$x(T)$ is unspecified, and that F in (2) is an explicit function of T, i.e. $F = F(x(T),T)$, then a necessary condition for the pair $(u^*(\cdot),T^*)$ to be optimal is that[†]

$$H(x^*(T^*), u^*(T^*), \lambda(T^*), \lambda_0, T^*) - \frac{dF}{dT}(x^*(T^*),T^*) = 0 \qquad (13)$$

2. EXPLOITATION OF A LUMPED RESOURCE. Consider the logistic growth model [6]

$$\frac{dx}{dt} = ax(1 - x/K) \qquad x(0) = x_0 , \qquad (14)$$

where x is a scalar population-density variable, $a > 0$ is the *intrinsic* per capita growth rate and $K > 0$ is the *environmental carrying capacity*, is perhaps the simplest model of population growth in a finite or saturating environment. The solution to (14) is the so-called *sigmoidal* or S-shaped curve

$$x(t) = K/(1 - ce^{-at}) , \qquad (15)$$

where c is a constant determined by x_0. Equation (14) has been used extensively to model the growth of single populations, including human populations [13]. More recently equation (14), has been used in the general framework of Problem 1 to analyze the exploitation of renewable resources such as fisheries and whale populations [10,11,14].

PROBLEM 2. (After Cliff and Vincent [14]). Maximize the objective functional

$$J_T(u(\cdot)) = \int_0^T (ux - \gamma u)\, dt \qquad (16)$$

over all piecewise continuous functions $u(\cdot): [0,T] \to [0,b] \subset R$ subject to

$$\frac{dx}{dt} = x(1-x) - ux \qquad x(0) = x_0 \qquad (17) \ \square$$

Here x and t are transformed to dimensionless units (cf. equation (14). The term ux is the harvest rate which is assumed to be proportional to the population density x and the effort level u. The parameter $\gamma > 0$ in (16) is a cost-price ratio. Thus J_T represents the net relative profit or rent (harvesting income versus effort expenditure) provided by the resource over the time interval $[0,T]$.

SOLUTION. Applying the maximum principle, bearing in mind that we are minimizing $-J_T$, it follows that the switching function (cf. equation (8)) is given by

$$\sigma(t) = x^*(t)(1 - \lambda(t)) - \gamma \qquad (18)$$

The problem is singular since its Hamiltonian (cf. equation (5)) is linear in u, i.e., u does not appear in (18). Thus the switching function cannot be used

[†]See previous footnote

directly to construct extremal controls. Setting $\sigma(t) = 0$ implies that along *singular arcs*, i.e. non bang-bang segments of the singular control, the relationship

$$\lambda(t) = 1 - \gamma/x(t) \tag{19}$$

holds. By differentiating both sides with respect to time and substituting for $\frac{dx}{dt}$ and $\frac{d\lambda}{dt}$ (cf. (6) and (17)) it can be shown that the singular arc is given by

$$\begin{aligned} x^s &= (1+\gamma)/2 \\ u^s &= (1-\gamma)/2 \\ \lambda^s &= (1-\gamma)/(1+\gamma) \end{aligned} \tag{20}$$

From (9), (10), (18) and (19) it follows that

$$u^*(t) = 0\ (u^s)\ (b) \quad \text{only if} \quad \lambda > (=)\ (<)\ \lambda^s \tag{21}$$

provided $u^s \in [0,b]$. Further, it is possible to show that $x_0 < (>)\ x^s \Rightarrow \lambda(0) > (<)\ \lambda^s$, i.e. there exists $t_1 > 0$ such that

$$x_0 < (>)\ x^s \quad \Rightarrow \quad u^*(t) = 0\ (b) \quad \text{for } t \in [0,t_1]. \tag{22}$$

Also, $x(T)$ unspecified implies that $\lambda(T) = 0$. Since (20) implies that $\lambda^s > 0$, it follows from (21) and $\lambda(T) = 0$ that there exists $t_2 < T$ such that

$$u^*(t) = b \quad \text{for} \quad t \in [t_2, T] \tag{23}$$

The values of t_1 and t_2 are obtained by solving the equation $\frac{dx}{dt}$ forward from $x(0) = x_0$ and the equation $\frac{d\lambda}{dt}$ backward from $\lambda(T) = 0$. Thus the extremal solution will include a singular component on $[t_1, t_2]$ only if T is sufficiently large to ensure that $t_1 < t_2$. Using sufficiency conditions [15], Cliff and Vincent [14] showed that the extremal solution determined by (22) and (23) is optimal. □

The singular arc specified by (20) is also the solution to the static problem (cf. (16) and (17))

$$\max_{u \geq 0} \quad ux - \gamma u \tag{24}$$

subject to

$$x(1-x) - ux = 0\ . \tag{25}$$

Since the static problem is equivalent to maximizing the rate of rent accumulation under an equilibrium constraint on (17), the solution to (24) and (25) is termed the *Maximum Sustainable Rent* (MSR) solution. The interpretation of the solution to Problem 2, from (22), is to approach the MSR as rapidly as possible and, from (23), to leave the MSR near the end of the planning horizon so that advantage can be taken of the lack of constraints or costs associated with leaving the resource depleted at time T. To avoid this artifact of a finite planning horizon $[0,T]$ several approaches can be taken. Allowing $T \to \infty$, presents the problem that $J_T \to \infty$. This can be avoided: by considering $\lim_{T\to\infty} J_T/T$ [16]; by introducing a discount factor $\delta > 0$ into the objective functional so that J_∞ is defined [10] i.e.

$$J_\infty = \int_0^\infty e^{-\delta t} (ux - \gamma u)\, dt; \tag{26}$$

or by deriving necessary conditions for Problem 1 on finite but arbitrarily large time intervals $[0,T]$ [17,18]. In all of these analyses, the MSR solution plays a central role. In the analysis of (26), the costate variable $\lambda(t)$ is replaced by the *current-value* costate variable $q(t) = e^{\delta t} \lambda(t)$.

The discussion so far has focused on harvesting a single logistic resource (i.e., a population modeled by [14]). The extension of Problem 2, to harvesting one or both populations in a general two-dimensional modeling framework, has been considered. Taking advantage of Poincare-Bendixon theory, Haurie [18] has analyzed the role of the MSR solution and extremal control. For problems linear in $u = (u_1, u_2)^T$, the MSR solution plays the role of an *exact turnpike* while, in general, the MSR solution is an *asymptotic turnpike*, i.e., all solutions respectively reach the MSR solution if T is sufficiently large, or asymptotically approach the MSR solution on infinite time intervals. For two competing populations, however, unlike the one-dimensional case, the optimal trajectory does not approach the MSR turnpike as rapidly as possible but approaches along *partially singular arcs* [19] (i.e., arcs along which one component of the control satisfies (9) and the other satisfies (10). The *minimum time* approach, however, involves only bang-bang controls (cf. (9)), i.e., $u(t) \in \{(a_1,a_2), (a_1,b_2), (b_1,a_2)\ (b_1,b_2)\}$, while extremal solutions to the harvesting problem also involve the elements $\{(a_1,u_2^s), (u_1^s,a_2), (u_1^s,u_2^s), (b_1,u_2^s), (u_1^s,b_2)\}$, assuming that the MSR solution satisfies $u_i^s \in (a_i,b_i)$ $i = 1,2$ (see [11] for further discussion).

3. MANAGEMENT OF PREY-PREDATOR INTERACTIONS. Volterra [7], modeled the cyclic behavior of a fish prey-predator system using two coupled quadratic differential equations. In dimensionless variables this model can be written as:

$$\begin{aligned} &\text{prey equation:} && \frac{dx_1}{dt} = x_1(1-x_2) \quad x_1(0) = x_{10} \\ &\text{predator equation:} && \frac{dx_2}{dt} = x_2(x_1-k) \quad x_2(0) = x_{20} \end{aligned} \tag{27}$$

Volterra showed that solutions to (27) are cycles whose amplitude depends on the initial values x_{i0}, $i = 1, 2$, i.e., system (27) possesses a constant of motion determined by

$$x_1(t)^k \, x_2(t) \, e^{-(x_1(t)+x_2(t))} = (x_{10})^k \, x_{20} \, e^{-(x_{10}+x_{20})}. \tag{28}$$

Goh et al. [20] considered the problem of driving the populations to the equilibrium solution (k, 1) while minimizing some performance criterion (objective functional). In a pest management context, the advantage of having the populations at equilibrium is the prevention of periodic outbreaks of the pest population x_1 beyond its equilibrium level k as it interacts with its natural enemy x_2.

PROBLEM 3 (After Goh et al. [20]). Minimize the objective functional

$$J(u(\cdot)) = \int_0^T (cx_1 + u)dt$$

over all piecewise functions $u(\cdot): [0,T] \to [0,b] \subset R^1$ subject to

$$\begin{aligned} \frac{dx_1}{dt} &= x_1(1-x_2) + e_1 ux_1 & x_1(0) &= x_{10} \\ \frac{dx_2}{dt} &= x_2(x_1-k) + e_2 ux_2 & x_2(0) &= x_{20} \end{aligned} \tag{29}$$

where u represents the rate at which a chemical that

(i) selectively kills the pest x_1 (i.e., $e_1 = -1$, $e_2 = 0$); or

(ii) selectively kills the predator x_2 (i.e., $e_1 = 0$, $e_2 = -1$); or

(iii) kills both x_1 and x_2 at rates $e_1 < 0$ and $e_2 < 0$,

is applied to the system or u represents the rate at which laboratory reared

(iv) pests x_1 (i.e., $e_1 = 1$, $e_2 = 0$), or

(v) predators x_2 (i.e., $e_1 = 0$, $e_2 = 1$) are released.

The constant $c > 0$ in $J(u)$ is the relative cost of the damage induced per unit pest. □

SOLUTION. Except in the case of (iii) the optimal solutions are purely bang-bang. In the case of (iii), however, singular subarcs can arise but their significance is not understood. In all but the latter case, the solutions can be characterized by constructing, in the phase plane, i.e., $\{(x_1,x_2)\}$, the switching curves that represent the set of points at which the bang-bang controls switch between $u^*(t) = 0$ and $u^*(t) = b$. The solution to (iv) is depicted in Figure 1. By following the trajectory from the initial condition, as indicated, we see that two separate pest release periods are required to drive the system to its equilibrium. The solution also demonstrates the notion of *pest control through pest release*, i.e., the prevention of a future pest outbreak by ensuring that the natural enemy population does not collapse when pest levels are low.

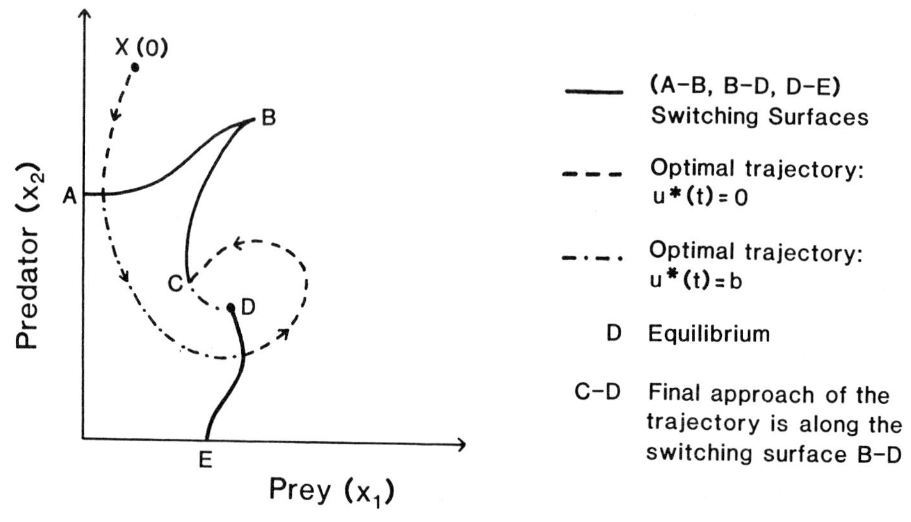

Fig. 1. Optimal stabilization of a pest and natural enemy interaction by releasing pests at the rate $u^* = b$ over, at most, two periods of time. (After Goh et al. [20]).

4. SEASONAL EXPLOITATION OF AN AGE-STRUCTURED FISHERY. A class of problems for which equation (3) does not directly provide an appropriate model, is the exploitation of age structured resources, such as fisheries, where each time period (year) is divided into a distinct spawning and harvesting season. In this case equation (3) holds during the harvesting season while a mapping is used to link the state variables between harvesting seasons.

PROBLEM 4 (After Getz [21]). Maximize the objective functional

$$J_N\,(u(\cdot)) = \sum_{k=0}^{N-1} \int_k^{k+t'} \sum_{i=1}^{n} w_i(t)\, u_i\, x_i\, dt \tag{30}$$

over all piecewise continuous functions

$u_i(\cdot): \bigcup_{k=0}^{N-1} [k,k+t'] \to [0,\infty)$, $i = 1, \ldots, n$, subject to the system constraint equations:

<u>Harvesting season</u>: $t \in [k, k+t']$, $k = 0, \ldots, N-1$ and $t' \in (0,1)$

$$\frac{dx_i}{dt} = -(\alpha_i + u_i)\, x_i \qquad i = 1, \ldots, n; \tag{31}$$

Spawning season: $k = 0, \ldots, N-1$

$$
\begin{aligned}
x_1(k+1) &= S\left(\sum_{i=1}^{n} c_i\, x_i(k+t')\right) \\
x_{i+1}(k+1) &= a_i\, x_i(k+t') \qquad i = 1, \ldots, n-2 \\
x_n(k+1) &= a_n\, x_n(k+t') + a_{n-1}\, x_{n-1}(k+t').
\end{aligned} \tag{32}
$$

□

In (31) the parameters α_i are natural mortality rates and the controls u_i are man-induced fishing mortality rates. Let $w(t)$ be a function describing the weight of an individual at time t that was born at $t = 0$. Then the weight of an individual in age class i at time $k + t$ is described by

$$w_i(t) = w(t - k + i - 1), \quad i = 1, \ldots, n \quad t \in [k, k + t']. \tag{33}$$

The integrand of (30) is thus the rate of biomass removal so that J_N represents the total biomass removed over the N-year exploitation period. As a weight function, $w(t)$ is assumed to be increasing with time, bounded above and its proportional rate of increase is assumed to decrease with time, i.e., there exists a constant $\bar{w} > 0$ such that

$$0 \leqslant w(t) \leqslant \bar{w}, \quad \frac{dw}{dt} < 0, \quad \frac{d}{dt}\left[\frac{dw}{dt}/w(t)\right] < 0. \tag{34}$$

In (32) the constants $a_i \in [0,1]$ represent the probability of surviving from age-class i at time $k + t'$ to age-class $i + 1$ at time $k + 1$. These constants are usually related to natural mortality rates α_i by

$$a_i = e^{-\alpha_i(1 - t')}$$

The constants $c_i \geqslant 0$ reflect the relative fecundity of individuals in the i-th age class so that

$$p = \sum_{i=1}^{n} c_i\, x_i(k + t') \tag{35}$$

can be regarded as a stock fecundity index. The function $S: R \to R$ is usually referred to as the stock-recruitment relationship.

Problem 4 is an optimal control with state variable jump discontinuities problem. In the calculus of variations this type of problem was analyzed by Denbow [22] and in an optimal control setting by Vincent and Mason [23]. A general maximum principle is stated in [24] and includes a transformation condition that the costate variable λ must satisfy across the jump interval $[k + t', k + 1]$. Specifically, if

$$x(k+1) = g(x(k + t')) \tag{36}$$

where $g: R^n \to R^n$ (cf. the particular case given by equation (32)), then the corresponding costate condition is

$$\lambda(k + t') = \frac{\partial g^T}{\partial x}(x(k+t'))\, \lambda(k + 1) \quad , \tag{37}$$

where $\frac{\partial g}{\partial x}$ is the gradient matrix associated with $g(x)$. A similar result also appears in [3].

In previous section, the importance of the MSR solution was demonstrated in terms of its *turnpike* property. Equations (31), however, do not possess a positive equilibrium solution, since for $u_i \geqslant 0$, $x_i(t)$ decays exponentially to zero. The transformation given by (32), however, can act to restore the density of the i-th age-class at time $k + 1$ to its level at time k as each age class moves one class ahead. In this case a *dynamic equilibrium* can occur that is characterized by

$$x_i(k + 1) = x_i(k) \qquad i = 1, \ldots, n. \tag{38}$$

If (38) holds for $k = 0, 1, 2, \ldots, N$ then Problem 4 reduces to solving for $u_i^*(\cdot)$, $i = 1, \ldots, n$ on $t \in [0, 1]$, i.e., from (30), $J_N(u^*(\cdot)) = NJ_1(u^*(\cdot))$

SOLUTION. Since the controls u_i appear linearly in Problem 4, extremal control consists of bang-bang and/or singular subarcs. For this problem it is easily shown that the time derivative of the switching function (cf. (8)) satisfies

$$\frac{d\sigma_i}{dt} = w_i(t)\, x_i^*(t)\left[\left(\frac{dw_i}{dt} / w_i(t)\right) - \alpha_i\right] \qquad i = 1, \ldots, n. \tag{39}$$

Since $w_i(t)$ and $x_i(t)$ are always positive, it follows from (33) and (34) that $\frac{d\sigma_i}{dt}$ can change sign at most once on $[0, t']$ and then only from negative to positive. Also, since $u_i(t) \geqslant 0$ is not bounded above, whenever extremal control is *switched on*, its value becomes infinite. Thus it is easily shown (see [21] for details) that extremal control is drawn from the class of *impulse functions* acting at time t_i^* where, from (39)

$$\begin{aligned} t_i^* &= 0 \quad \text{if} \quad \left[\frac{dw_i(0)}{dt} / w_i(0)\right] \leqslant \alpha_i \\ t_i^* &= t' \quad \text{if} \quad \left[\frac{dw_i(t')}{dt} / w_i(t')\right] \geqslant \alpha_i \end{aligned} \tag{40}$$

or if neither condition in (40) holds then $t_i^* \in (0, t')$ and is the solution to the equation

$$\frac{dw_i(t_i^*)}{dt} / w_i(t_i^*) = \alpha_i \tag{41}$$

Applying an impulse control at time t_i^* is equivalent to removing a given number of individuals, say z_i, at that point in time. In this case it can be shown that the solution to Problem 4 subject to constraints (38) can be obtained by solving the following parametric linear programming problem (see [25] for details):

maximize

$$J_1 = \sum_{i=1}^{n} w_i(t_i^*) z_i \tag{42}$$

over the set $\{p, x_1 > 0, z_i \geqslant 0 \quad i = 1, \ldots, n\}$ subject to the constraints

$$r_0 x_1 - \sum_{i=1}^{n} r_i z_i = p \tag{43}$$

$$x_1 = S(p) \tag{44}$$

$$\sum_{i=1}^{n} q_i z_i - x_1 \leqslant 0 \tag{45}$$

The objective function (42) is equivalent to J_1 as defined for $N = 1$ in (30). S is the stock recruitment function introduced in (32) and p is the stock parameter defined by (35). The parameters q_i can be expressed in terms of the parameters α_i and a_i in (31) and (32) and the inequality (45) is an expression of the fact that under equilibrium constraints (38) the number of individuals removed from the population cannot exceed the number x_1 that are recruited. Equation (43) is derived from (35) so that in addition to involving q_j and α_j, $j = 1, \ldots, n$, the parameters r_i, $i = 1, \ldots, n$, include the constants c_j, $j = 1, \ldots, n$. Finally, we can use the fundamental theorem of linear programming, which asserts that if an optimal feasible solution exists a basic optimal feasible solution exists [26], to identify the nature of the solution to Problem 4. For the problem defined by (42)-(45), a basic solution involves at most $x_1 > 0$ and two elements of $z = (z_1, \ldots, z_n)^T$. Taking cognizance of which constraints are tight (i.e., hold with equality) the following theorem can be stated [25]

THEOREM. The optimal solution to Problem 4 involves the application of impulse control to at most two age classes, x_{i_1} and x_{i_2} say, at times $t_{i_1}^*$ and $t_{i_2}^*$ determined from (40) and (41). Further, if two age classes are harvested, it is optimal to remove all individuals from the older class. □

The *bimodal nature of the optimal harvesting policy* is independent of the form of S. This bimodal result was first obtained for linear S by Beddington and Taylor [27]. More recently Feichtinger [28] has shown the result to hold for an infinite dimensional problem dealing with continuous age distributions.

5. EXPLOITATION AND INVESTMENT. In Problem 2 it was assumed that the maximum level of exploitation was fixed at a predetermined level b. In a fishery, for example, this would be related to the maximum number of boats available to exploit the resource. Also, in Problem 2, economic considerations were confined to considering a cost-price ratio γ. The economic aspects of this problem have been examined in much greater detail [10,29,30]. The problem presented below examines the role of investment in maximizing the *present value* of all future rent derived from exploiting a resource.

PROBLEM 5 (After Clark et al. [29]). Maximize the objective functional

$$J(u(\cdot),\ v(\cdot)) = \int_0^\infty e^{-\delta t}\ [ux - \gamma u - \pi v]\ dt \tag{46}$$

over all piecewise continuous functions $u(\cdot):[0,\infty) \to [0,b(t)] \subset R$ and $v(\cdot):[0,\infty) \to V \subseteq R$ subject to

$$\frac{dx}{dt} = f(x) - ux, \qquad x(0) = x_0 \tag{47}$$

and

$$\frac{db}{dt} = -\beta b + v, \quad b(0) = b_0 \tag{48}$$ □

As in Problem 2 the control function $u(t) \leqslant b(t)$ is the exploitation effort level and ux is the harvest rate. The maximum effort level b(t), however, is now a dynamic variable that reflects the level of capitalization in the industry. From (48), this capital depreciates at a rate $\beta > 0$, but can be supplemented through investment at a rate determined by the control variable v. As before, $\gamma > 0$ is the cost-price ratio, while the parameter $\pi > 0$ represents the relative cost of purchasing one unit of b. The objective functional (46) is expressed in terms of the present value of the resource, since it includes a discount factor $\delta > 0$ on all future revenues (cf. expression (26)).

SOLUTION. Problem 5 departs from the previous problems in that the control constraint set for $u(\cdot)$ is now dependent on a state variable, i.e., b(t) satisfying (48). The Maximum Principle does not hold for state-dependent control constraints and a more general set of necessary conditions are needed [4,31]. For the special case V = R, i.e., capital can be freely shifted in and out of the resource, the optimal policy will be $u^*(t) = b(t)$ since the excess units of capital can be directly converted into cash. In a fishery, this would correspond to a pool of fishing vessels that can be transferred between different fisheries, as the demand arises. Integrating by parts, utilizing (48), the functional (46) reduces to

$$J(u(\cdot)) = \int_0^\infty e^{-\delta t} [x - \gamma - \pi (\delta+\beta)] u \, dt \tag{49}$$

Thus, for the case $V = R$, Problem 5 reduces to maximizing (49) over all piecewise continuous $u(\cdot)$: $[0,\infty) \to [0,\infty)$ subject to (47). The problem now falls within the scope of the Maximum Principle and has the classical *turnpike* solution (cf. section 2), i.e., the resource is driven as rapidly as possible (in this case an ad hoc upper bound or $u(t)$ must be imposed if an impulse solution is considered unacceptable) to the singular MSR solution, x^s, which satisfies [10,29,30]

$$f'(x^s) - [\phi'(x^s)f(x^s)/(1 - \phi(x^s))] = \delta \tag{50}$$

where

$$\phi(x) = [\gamma + \pi(\delta + \beta)]/x \tag{51}$$

and $'$ denotes differentiation with respect to x. Note that x^s in (20) is the solution to (50) for the case $\pi = \delta = 0$ and $f(x) = x(1-x)$. Now consider the case $V = [0,\infty)$, i.e., it is not possible to withdraw capital from the fishery. If the stock is initially in a virgin state, then it has been shown [29] that the initial level of capitalization always exceeds the long run optimum b^s (cf. u^s defined in (20)). The optimal solution thus includes initially driving the stock level to below the MSR solution x^s before depreciation of $b^*(t)$ allows $x^*(t)$ to approach x^s, at which point an optimal investment rate $v^s = \beta b^s$ (cf. (48)) is established to maintain the resource at its MSR level. □

In the fisheries literature excess capitalization has often been regarded as a sign of poor management. The above theory indicates, however, that in the case where the capital is *non-malleable*, i.e., $V = [0,\infty)$, the optimal solution includes an initial *over-capitalization* phase.

6. PESTICIDES AND RESISTANCE. In the use of pesticides a new consideration enters into the analysis of population management, that is, the influence of the control variables with time may be reduced as the population becomes resistant to their use. In most populations, the response of individuals to the application of pesticides will vary, according to their degree of susceptibility. The pesticide thus provides a selective force that favors the survival and reproduction of the most resistant individuals in the population. With each application of the pesticide the proportion of resistant individuals in the subsequent generation is enhanced, often to the point where further application of pesticide is ineffectual. This problem has been analyzed in several

settings [32-35]. Here we present a study that considers the problem of applying *alternative technologies*, i.e., the problem of when it is optimal to switch management practices such as switching from one pesticide to another, switching to natural control, or planting a different crop.

PROBLEM 6 (After Regev et al. [33]). Maximize the objective functional

$$J = \int_0^T e^{-\delta t} [p(x) - cu] \, dt + F_i(x(T), w(T)) \, e^{-\delta T} \tag{52}$$

over all piecewise continuous $u(\cdot): [0,\infty) \to [0,\infty)$, all $i \in I$ (a finite index set) and all $T \geqslant 0$ subject to

$$\frac{dx}{dt} = x(1 - k(w,u)) \, f(x) \qquad x(0) = x_0 \tag{53}$$

$$\frac{dw}{dt} = w \, g(w,u) \qquad w(0) = w_0. \tag{54}$$

□

The variables $x(t)$ and $w(t)$ respectively denote the level of pests and the proportion of individuals in the population that are resistant to a pesticide applied at a rate $u(t)$. The per capita growth rate $f(x)$ is specified as satisfying

$$\frac{\partial f}{\partial x} < 0 \quad \text{and} \quad f(\bar{x}) = 0 \quad \text{for some} \quad \bar{x} > 0. \tag{55}$$

The proportional rate of increase of resistant individuals is specified as satisfying

$$\frac{\partial g}{\partial u} > 0 \ , \ \frac{\partial g}{\partial w} > 0 \ \text{ and } \ g(1,u) = 0 \ \text{ for all } u > 0. \tag{56}$$

These conditions, together with (54), imply that the level of resistance $w(t)$ can only increase with time and that $w(t) \to 1$ as long as $u(t) > 0$. The kill rate function $k(w,u)$ is specified, for all $w(t) \in [0,1]$, as satisfying

$$\begin{gathered} k(w,0) = 0 \ \text{ and } \ k(w,u) \in [0,1] \text{ for } u \geqslant 0 \ , \\ \frac{\partial k}{\partial w} < 0 \ \text{ and } \ \frac{\partial k}{\partial u} > 0 \ . \end{gathered} \tag{57}$$

In the objective functional (52), the function $p(x) > 0$ represents the value of the output from the crop. This value is assumed to decrease with increasing x, i.e., $\frac{\partial p}{\partial x} < 0$. The parameter $c > 0$ is the cost of applying the pesticide and $\delta > 0$ is the discount rate (cf. Problem 5). The functions $F_i(x(T),w(T))$, $i \in I$, represent the profits, discounted on $[T,\infty)$, that can be realized by switching to alternative technologies and are assumed to satisfy

$$\frac{\partial F_i}{\partial x} \leqslant 0 \quad \text{and} \quad \frac{\partial F_i}{\partial w} \leqslant 0 \tag{58}$$

SOLUTION. Application of the Maximum Principle is straightforward. In this case the value of T is unspecified and the optimal value satisfies (13). Since for any $i \in I$,

$$\frac{d}{dT}[F_i(x,w)e^{-\delta T}] = -\delta F_i(x,w) \tag{59}$$

the interpretation of (13) for this problem is: the shift to the i-th technology will occur only when the value of the Hamiltonian along the optimal trajectory is equal to the value of (59), interpreted as the *instantaneous future gain*, along the optimal trajectory. As a consequence of its properties, as specified in (58), an alternative technology F_i will never be implemented if (cf. (55) for a definition of $\overline{x}$).

$$p(\overline{x}) - cu > \delta F_i(\overline{x}, 1), \tag{60}$$

i.e., an alternative technology cannot be feasible if its instantaneous future gain rate under maximum *resistant* pest levels is less than the instantaneous return obtained at maximum pest levels using the current technology. Thus inequality (60) allows policy makers to exclude all non-profitable alternatives. □

Regev et al. [33], also compared the solution to Problem 6 with the solutions to two related problems, viz, the problem where resistance is set at w_0 and the dynamics of $w(t)$ are ignored, and the so-called competitive solution where resistance is regarded as an unmanaged *common property* resource. Their analysis leads to the conclusion that the competitive solution is often preferable to a centralized optimal management plan that ignores an increase in resistance with time.

7. ALLOCATION OF RESOURCES. The problem of maximizing over a fixed time period the total output from a resource, where the rate of output is dependent on the size of the resource and the growth of the resource can only take place at the expense of producing output, has many applications. In population biology, the evolution of life history strategies in organisms, such as annual plants [36] and honey bee colonies [37] that allocate resources between *vegetative growth* (leaves, stems, worker bees, etc.) and *reproductive organs* (fruit, seeds, drones and queens, etc.) falls within the framework of this general problem. The same is true for the production of lymphocytes in the mammalian immune response system [38]. Although these specific

problems do not deal with an externally imposed management plan but are, rather, cybernetic in nature, their analysis yields principles of interest to population managers. Here we present the problem of maximizing the annual seed production of a plant that lives for N years.

PROBLEM 7 (After Mirmirani and Oster [39]). Maximize the objective functional

$$J = \sum_{i=1}^{N} p^{i-1} s_i(i) \tag{61}$$

over all piecewise continuous functions $u(\cdot):[0,N] \to [0,1]$ subject to the system constraints

$$\frac{dx}{dt} = aux \qquad x(0) = x_0 \tag{62}$$

$$\frac{ds_i}{dt} = b(1-u)x \qquad \text{for } t \in [i-1, i) \text{ and } s_i(i-1) = 0 \qquad i = 1, \ldots, N. \tag{63}$$

□

The variable $x(t)$ represents the biomass of the plant as it grows over the N year period while the variables s_i, $i = 1, \ldots, N$, represent the biomass of seeds produced by the plant during its i-th year. The control variable $u(t) \in [0,1]$ represents the proportion of photosynthate allocated to growth versus seed production at time t. The constants $a > 0$ and $b > 0$ respectively reflect the relative efficiences in converting photosynthate to plant and seed growth at a rate proportional to the plant biomass x. The probability of the plant surviving from one year to the next is assumed to be constant and denoted by $p \in (0,1]$. As such it is regarded as a discount factor.

SOLUTION. Like Problem 4, Problem 7 can be considered in the formulation of control systems with state variable jump-discontinuities. The solution to Problem 7, however, can be explicitly derived, since x and u appear linearly in equations (61)-(63) and $s_i(t)$ does not enter the dynamics. Using the Maximum Principle it is easily shown that extremal control is bang-bang, i.e., singular subarcs do not exist, and has the form

$$\begin{aligned} u^*(t) &= 1 \qquad t \in [i-1, t_i^*) \qquad i = 1, \ldots N \\ u^*(t) &= 0 \qquad t \in [t_i^*, i) \end{aligned} \tag{64}$$

The values of t_i^* can be characterized as follows. Define

$$\tau_i = i - (1/a)(1 - \underbrace{pe^{a-(1-pe^{a-(\ldots)})}}_{N-i \text{ terms}}) \qquad i = 1, \ldots N \tag{65}$$

where $a > 0$ is the growth rate in (62), i.e. $\tau_N = N - 1/a$ and $\tau_{N-1} = N-1 - (1/a)(1-pe^{a-1})$ etc. Then

$$\begin{aligned} &t_i^* = i \quad \text{if} \quad \tau_i \geqslant i \\ \text{or} \quad &t_i^* = i-1 \ \text{if} \quad \tau_i \leqslant i-1 \\ \text{or} \quad &t_i^* = \tau_i \quad \text{if} \quad \tau_i \in (i-1, i). \end{aligned} \tag{66}$$

If the first case in (66) holds then it follows from (62)-(64) that all resources are allocated to growth during the i-th season (year). If the second case in (66) holds then it follows that all resources are allocated to seed production during the i-th season. If $\tau_i \in (i-1,i)$, however, then it follows from (64) and (66) that the plant switches from growth to seed production at time τ_i. From (65) it is easily shown that

$$i - \tau_i > (i-1) - \tau_{i-1} \qquad i = 2, \ldots, N,$$

so that once seed production begins then in each subsequent year the onset of seed production occurs earlier in the season. The final season will be solely devoted to seed production only if $a < 1$. This follows since $\tau_N = N-(1/a)$. It follows from (65) that seed production during the early seasons will occur if p is sufficiently small, i.e., the plant will start producing seeds early in its N year life cycle if it has a low probability of surviving from one year to the next. In fact, it has been shown [39] that, no matter how large N is, if a plant is reproducing optimally, it will produce seeds during its first season whenever $p < e^{-a}$. Finally, for finite N it can be demonstrated that the extremal control defined in (64)-(66) is optimal (cf. [38]). □

Recently Problem 7 was generalized by Schaffer et al. [40] to include a storage component thus allowing photosynthate to be either allocated immediately towards vegetative growth or seed production or to be stored for future use. Given that a plant's photosynthetic rate and reproductive efficiency vary through the season, with the former peaking earlier in the season than the latter, conditions are presented in [40] which favor the evolution of a storage component in a plant.

8. CONCLUSION. In each of Sections 2-7 a problem and the essence of its solution are presented. Although these solutions suggest principles relating to the management and growth of populations, the application of these principles, both to infer understanding and to facilitate the design of management programs must be approached with caution. Firstly, the population models, are highly simplified abstractions of real world processes. Secondly, the framework

in which the management problems are formulated as idealizations of a complex set of biological, economic and sociological issues. Nevertheless, these principles do help focus our analysis of the problems at hand. The challenge we face is to integrate the insights acquired from *analyzing Platonic formulations* with the knowledge gained from many years of *seat of the pants practical experience* so that better resource management decisions can be made in the future.

BIBLIOGRAPHY

1. L. S. Pontryagin, V. G. Boltyanskii, R. V. Gamkrelidze and E. F. Mishchenko, *The Mathematical Theory of Optimal Processes*, Interscience Publishers, New York, 1962.

2. M. Athans and P. L. Falb, *Optimal Control*, McGraw-Hill, New York, 1966.

3. A. E. Bryson and Y. C. Ho, *Applied Optimal Control*, Blaisdell, Lexington, Massachusetts, 1969.

4. G. Leitmann, *The Calculus of Variations and Optimal Control*, Plenum, New York, 1981.

5. B. Gompertz, *On the nature of the function expressive of the law of human mortality*, Philos. Trans. R. Soc., London 115 (1825), 513-585.

6. P. F. Verhulst, *Notice sur la loi que la population suit dans son acroissement*, Corresp. Math. Phys. 10 (1838), 113-126.

7. V. Volterra, *Variazione e fluttuazini del numero d'individui in specie animali conviventi*, Mem. Acad. Naz. Linai (Ser 6) 2 (1923), 31-113.

8. A. J. Lotka, *Elements of Mathematical Biology*, Dover, New York, 1956 (reprint of 1925 edition).

9. G. F. Gause, *The Struggle for Existence*, Williams and Wilkins, Maryland, 1934.

10. C. W. Clark, *Mathematical Bioeconomics: The Optimal Management of Renewable Resources*, Wiley Interscience, New York, 1976.

11. B. S. Goh, *Management and Analysis of Biological Populations*, Elsevier, Amsterdam, 1980.

12. D. J. Bell and D. H. Jacobson, *Singular Optimal Control Problems*, Academic Press, New York, 1975.

13. R. Pearl, *Studies in Human Biology*, Williams and Wilkins, Maryland, 1924.

14. E. M. Cliff and T. L. Vincent, *An optimal policy for a fish harvest*, J. Optim. Th. Appl. 12 (1973), 485-496.

15. G. Leitmann, *Sufficiency theorems for optimal control*, J. Optim. Th. Appl. 2 (1968), 285-292.

16. T. L. Vincent, C. S. Lee and B. S. Goh, *Control targets for the management of biological systems*, Ecol. Model. 3(1977), 285-300.

17. H. Halkin, *Necessary conditions for optimal control with infinite horizon*, Econometrica 42(1974), 267-273.

18. A. Haurie, *Stability and optimal exploitation over an infinite time horizon of interacting populations*, Opt. Control Appl. Meth. 3 (1982), 241-256.

19. W. M. Getz, *On harvesting two competing populations,* J. Optim. Th. Appl. 28(1979), 585-601.

20. B. S. Goh, G. Leitmann, T. L. Vincent, *The optimal control of a prey predator system,* Math Biosci. 19(1974), 263-286.

21. W. M. Getz, *Optimal harvesting of structured populations,* Math Biosci. 44 (1979), 269-291.

22. D. H. Denbow, *A Generalized Form of the Problem of Bolza,* Ph.D. Thesis, University of Chicago, 1937.

23. T. L. Vincent and J. D. Mason, *Disconnected optimal trajectories,* J. Optim. Th. Appl. 3(1969), 263-281.

24. W. M. Getz and D. H. Martin, *Optimal control theory with state variable jump discontinuities,* J. Optim. Th. Appl. 31 (1980), 195-205.

25. W. M. Getz, *The ultimate-sustainable-yield problem in nonlinear age-structured populations,* Math. Biosci. 48(1980), 279-292.

26. D. G. Luenberger, *Introduction to Linear and Nonlinear Programming,* Addison-Wesley, Reading, Massachusetts, 1973.

27. J. R. Beddington and D. B. Taylor, *Optimum age specific harvesting of a population,* Biometrics 29(1973), 801-809.

28. G. Feichtinger, *Optimal bimodal harvest policies in age-specific bioeconomic models.* In, G. Feichtinger and P. Kall (eds.), Operations Research in Progress, 285-299, Reidel Publ. Co., 1982.

29. C. W. Clark, F. H. Clark and G. R. Munro, *The optimal exploitation of renewable resource stocks: problems of irreversible investment,* Econometrica 47 (1979), 25-47.

30. C. W. Clark, *Mathematical models in the economics of renewable resources*, SIAM Review 21 (1979) 81-99.

31. F. H. Clarke, *Necessary conditions for a general control problem,* In, D. L. Russell (ed.), Calculus of Variations and Control Theory, Academic Press, New York, 1976.

32. D. Heuth and U. Regev, *Optimal agricultural pest management with increasing pest resistance,* Amer. J. Agri. Econ. 56 (1974), 543-552.

33. U. Regev, H. Shalit, A. P. Gutierrez, *On the optimal allocation of pesticides with increasing resistance: the case of alfalfa weevil,* J. Environ. Econ. Manage. (to appear).

34. R. E. Plant and M. Mangel, *Multiseasonal management of an agricultural pest II: the economic optimization problem,* Dept. of Mathematics, University of California, Davis, 1982 (preprint).

35. C. A. Shoemaker, *Optimal integrated control of pest populations with age structure,* Oper. Res. 30 (1982), 40-61.

36. T. L. Vincent and H. R. Pulliam, *Evolution of life history strategies for an asexual annual plant model,* Theor. Pop. Biol. 17 (1980), 215-231.

37. S. Macevic and G. Oster, *Modeling social insect populations II: optimal reproductive strategies in annual eusocial insect colonies,* Behav. Ecol. Sociobiol. 1 (1976), 265-282.

38. A. Perelson, M. Mirmirani and G. Oster, *Optimal strategies in immunology I. B-cell differentiation and proliferation,* J. Math. Biol. 3 (1976), 325-367.

39. M. Mirmirani and G. Oster, *Competition, kin selection and evolutionary stable strategies,* Theor. Pop. Biol. 13 (1978), 304-338.

40. W. M. Schaffer, R. S. Inouye, T. S. Whittam, *Energy allocation by an annual plant when the effects of seasonality on growth and reproduction are decoupled*, Am. Nat. 120 (1982), 787-815.

Division of Biological Control
University of California at Berkeley
1050 San Pablo Avenue
Albany, CA 94706

Proceedings of Symposia in Applied Mathematics
Volume 30, 1984

Graph Theory, Homology and Food Webs

George Sugihara[1]

1. INTRODUCTION

Ecological systems consist of large networks of interrelated parts. The living components of these networks are linked largely by trophic interactions involving predator-prey encounters and competition.

Our aim here will be to show how the formal tools of graph theory and algebraic topology may be brought to bear to uncover patterns in the structure of real food webs. We will begin by examining the various ways that food webs can be represented, and will then apply these representations to search for regularities in the structure of natural ecosystems.

The pattern of interactions in food webs is of scientific interest because it determines the flow of energy and materials (e.g., nutrients and pollutants) through ecological systems as well as the dynamics and stability of species populations. Knowledge that real ecosystems are contained within a specific subset of possible topologies has fundamental interest as it helps to illuminate those narrow regions in a wide landscape of mathematical possibilities that warrant further study.

2. CHARACTERIZING FOOD WEB STRUCTURE

The basic information that is used to construct a food web is contained in the so-called *food web matrix A* (Fig. 1). This is a binary matrix whose columns C correspond to the set of consumer species and whose rows R correspond to the set of resources or prey in the system. The l's in the matrix signify

[1]Research supported by an Ogden Porter Jacobus Prize from Princeton University and a Eugene P. Wigner Fellowship from Oak Ridge National Laboratory (operated by Union Carbide Corporation under contract W-7405-eng-26 with the U.S. Department of Energy), Oak Ridge, Tennessee. Publication No. 2229, Environmental Sciences Division, ORNL.

that a given consumer uses a given prey item which, in turn, generates the ordered pairs $(C_i R_i)$. Notice that a given column C_i in A identifies the subset of resources used by that consumer and each row R_i identifies the subset of consumers that share a given resource.

Ordered pairs in the food web matrix can be put into graphical form by constructing the *digraph* of trophic flows (Fig. 2). The digraph $D(V,E^*)$ consists of a finite nonempty set V of vertices formed from the union of consumers C and resources R, along with an arc set E^* containing the ordered pairs (C,R) from V. The vertices V in $D(V,E^*)$, therefore, consist of the set of all species in the food web (both predators and prey) and the directed edges E^* indicate who eats whom. The digraph is the classical standard for representing food webs (e.g., see Shelford 1913).

Two newer representations involving undirected graphs (no ordered pairs) may be constructed by choosing either consumers or resources, separately, as

Community Food Web Matrix (Knysna Estuary)

	12	13	14	15
2			1	
5				1
6	1			1
7	1	1		
8		1	1	
9		1	1	
10		1		
11		1	1	
13				1
14				1

LEGEND

2. attached plants
5. Hyporhamphus
6. Mugil
7. Upogebia
8. Lamya
9. Solen
10. Arenicola
11. Hymenosoma
12. Johnius
13. Lithognathus
14. Rhabdosargus
15. Hypacanthus

FIGURE 1. Data from Day (1967) (Sugihara 1982).

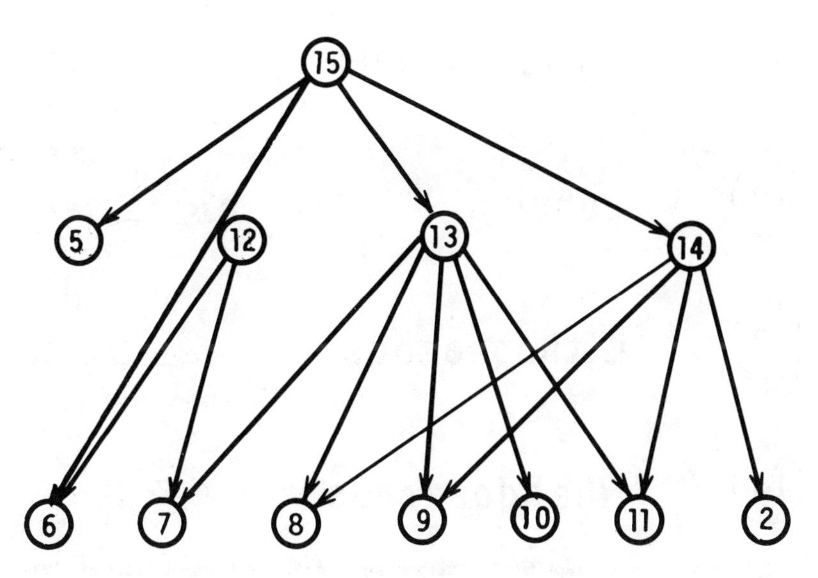

FIGURE 2. Digraph for the Knysna Esturary community (Day 1967) (Sugihara 1982).

object sets. These undirected graphs are examples of *intersection graphs* defined as follows.

Let $S = (S_1, \ldots, S_p)$ be an object set consisting of a nonempty family of distinct subsets. The intersection graph $G(S)$ is formed by identifying S with the vertices of G and making S_i and S_j adjacent whenever $i \neq j$ and $S_i \cap S_j \neq 0$. Hence an edge is drawn between two vertices in $G(S)$ whenever the subsets that they represent overlap. Marczewski (1945) has shown that every graph is an intersection graph.

The intersection graph that is formed by choosing $C = (C_1, \ldots, C_p)$ as the object set is Cohen's (1979) *consumer overlap graph G(C)* (Fig. 3). Here, whenever two consumers (subsets of resources) C_i and C_j overlap with respect to at least one resource an edge is drawn between them. The vertices in $G(C)$ represent consumer species and the edges signify overlap in consumer resource use. Such overlap may be determined mechanically by checking the columns of A for nonzero dot products.

The consumer overlap graph provides a portrait of the potential competitive structure of a food web. It is a simplified picture, however, in that a single edge joining two consumer species may represent overlap with respect to multiple resources.

A fuller characterization of consumer interrelationships may be obtained by taking the formal conjugate of $G(C)$ (Sugihara 1982). Herein let R be the object set and form the intersection graph $G(R)$ over R by joining those

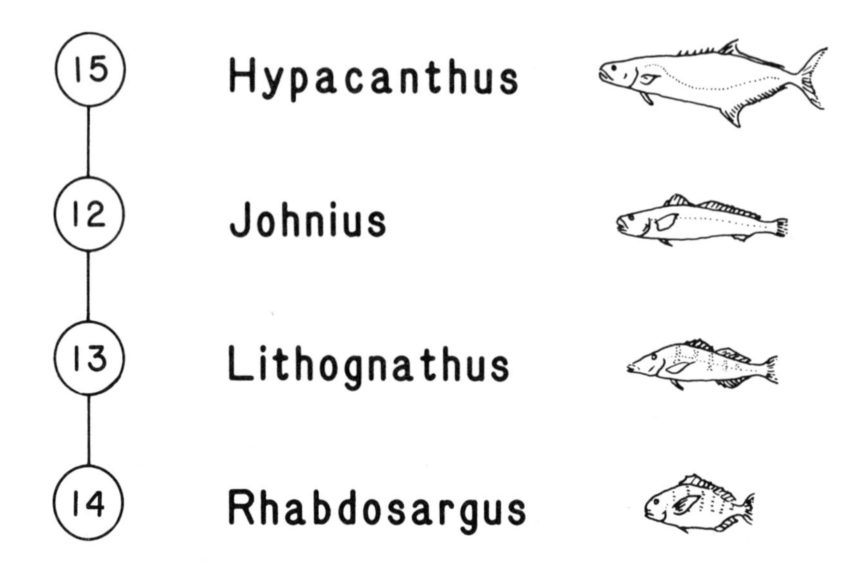

FIGURE 3. Niche overlap graph $G(C)$ corresponding to matrix in Figure 1 (Sugihara 1982).

resources that belong to a single consumer species. This involves checking for nonzero dot products between the rows of A.

The resulting *resource graph G(R)* (Fig. 4) gives a more complete picture than $G(C)$ for two reasons. First, whereas a consumer species in $G(C)$ is represented as a single vertex, in $G(R)$ it is represented by the constellation of resources used. That is to say, the *n-dimensional trophic niche* of a species in $G(C)$ corresponds to a *clique* (maximally connected subgraph) over its n resource vertices. A species that uses 5 resources is represented in $G(R)$ by its 5-dimensional niche, that is, as a clique over its 5 labeled resource vertices (see shaded area in Fig. 4). Secondly, unlike $G(C)$, $G(R)$ contains information as to the number of resources involved in the niche overlaps. As can be seen in Figure 4 the intersection of species niches may involve multiple resources.

For pedagogical purposes it may be useful to construct the multigraph (a graph having multiple edges) version of $G(R)$ out of tinker toys. Balls of modelling clay may be used to represent resource vertices, and an individual species niche may be represented by struts of a given color maximally connecting the resource set. Such a model is always possible due to the fact that any graph can be imbedded (with no edges crossing) in 3-space using only straight lines (e.g., Boesch 1981).

The information in the resource graph $G(R)$ may be given a more robust geometry if the notions of clique and graph are replaced with the terms simplex and complex as defined below.

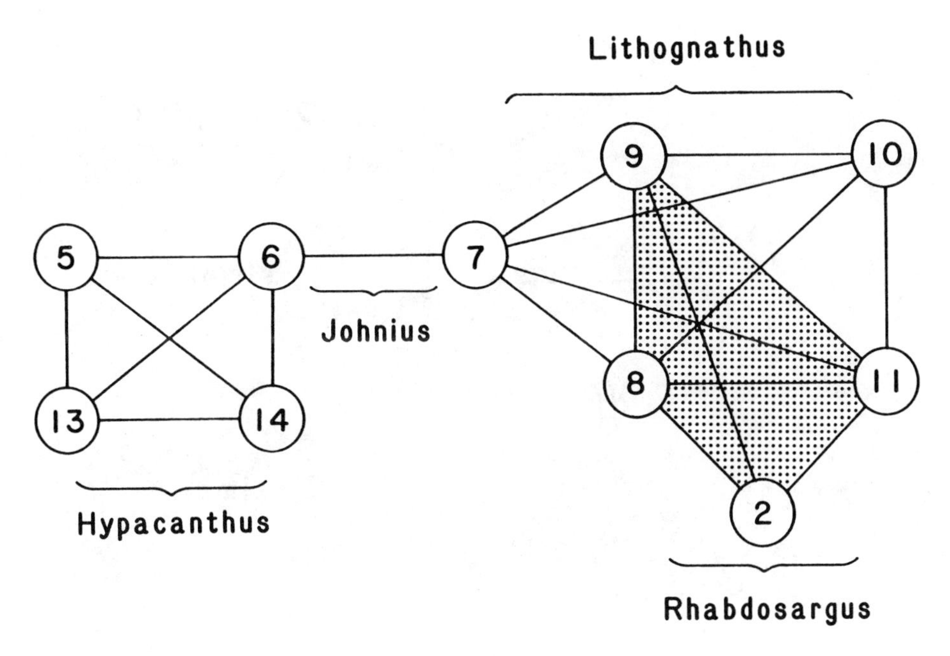

FIGURE 4. Resource graph $G(R)$ corresponding to matrix in Figure 1 (Sugihara 1982).

Let $v^o, \ldots, v^n$ be vectors in R_n. A vector V is said to be *affinely dependent* on them if there exist real numbers $\lambda_0, \ldots, \lambda_n$ called *barycentric coordinates* such that

$$\sum_{i=0}^{n} \lambda_i = 1 \quad \text{and} \quad V = \lambda_o v^o + \ldots + \lambda_n v^n \ .$$

Suppose further that $v^o, \ldots, v^n$ are affinely independent (none affinely dependent on the rest). Then, a *simplex* (closed) having vertices $v^o, \ldots, v^n$ is defined to be the set of points affinely dependent on $v^0, \ldots, v^n$ such that every barycentric coordinate is $\geqslant 0$. The *boundary* of a simplex consists of those points which have at least one barycentric coordinate equal to 0, and its *dimension n* is simply the number of vertices minus 1.

Notice how the notion of simplex corresponds with the graphical term "*clique*". Both are similar in that vertices and edges in the simplex identify with vertices and edges in the clique, however in dimensions above 1 the simplex characteristics of interior, exterior and boundary, as defined by the barycentric coordinates, differ from those in the graph. Harary (1972) defines a graph as a complex of simplexes of dimension 0 or 1. A simplex, therefore, can be viewed

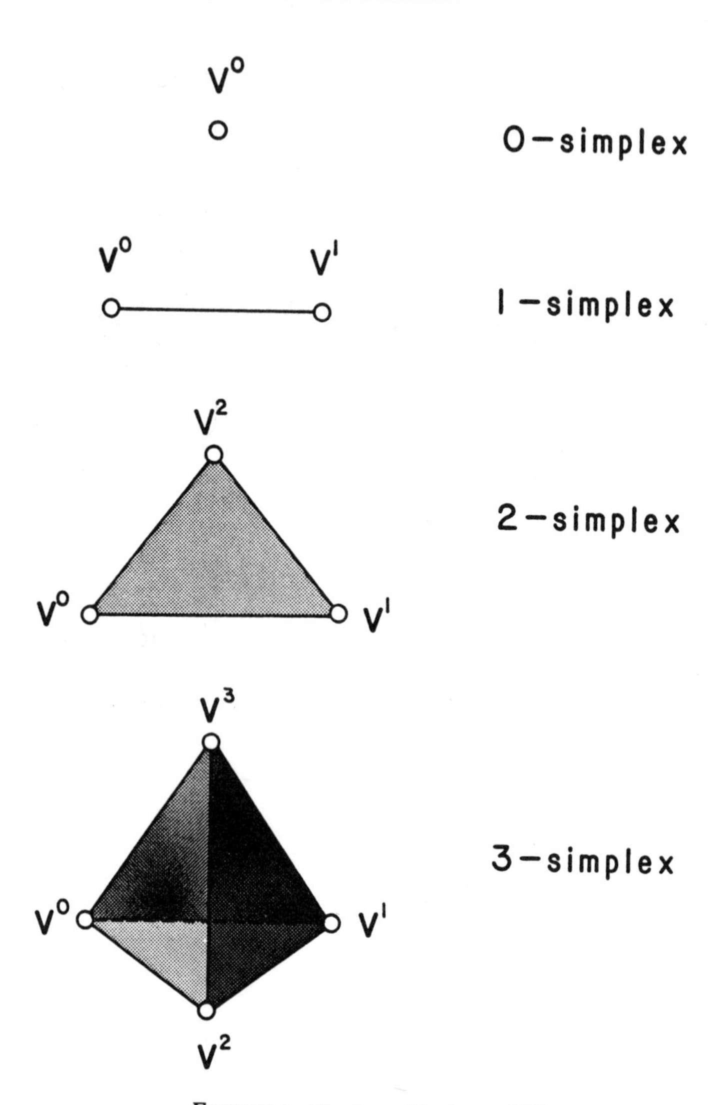

FIGURE 5. Simplexes (Sugihara 1982).

as a clique which has been inflated in a higher dimensional space to have an interior that is filled, and a clique conversely can be viewed as the vertices and edges of a simplex, or the *1-skeleton* of a simplex. Notice that because a clique is convex, so is the simplex. That is, if u and w are two vertices of the simplex then every straight line segment joining u and w belongs to the simplex. Indeed the simplex is the smallest convex set containing its vertices which is expressed by saying it is the *convex hull* of its vertices.

Paired with our graphical representation of the species niche in $G(R)$ as an n-pointed clique; it is also possible to represent the n-dimensional niche of a species as a simplex over n resource vertices. In geometrical terms this gives rise to a solid tinker toy model where a species niche may be thought of as a convex polyhedron whose n vertices correspond to the n resources used by that species. This generates a powerful architectural description of communities where the polyhedral species niches are joined through shared resource vertices to form a *simplicial complex* $K(R)$ (Fig. 6).

According to this model a community hypervolume may be broadly visualized as a multidimensional crystal having clusters of crystal faces which correspond to individual species niches. The simplicial complex model allows one to visualize clearly the geometry by which species niches are fitted together in communities.

3. TOPOLOGY OF REAL FOOD WEBS

We shall now apply these characterizations to a collection of 40 community food web matrices recently extracted from the literature (Sugihara 1982, after Briand 1983). In all, this collection contains 73 *nontrivial communities*; defined here as *connected components* (all vertices linked through some sequence of edges) in $G(C)$ containing 2 or more consumer species.

3.1 *HOLES: THE GEOMETRY OF SPECIES PACKING*

A question of biological interest concerns how closely species niches are packed together in $K(R)$. That is, are species niches packed tightly so that each community in $K(R)$ is a simple solid (Fig. 6B), or are they fitted together more loosely so that $K(R)$ resembles a multidimensional swiss cheese (Fig. 6A)? In biological terms, we are asking whether there are any minimal constraints acting on the geometry of species packing.

3.1.1 *TECHNICAL PRELIMINARIES*

The mathematics which we will use to address this question is *homology theory* or the study of multidimensional holes. The full technical details of these ideas are rather baroque, and the interested reader is refered to the excellent expositions given by Atkin (1974) and Giblin (1977). It will suffice to say, however, that basic information as of the holiness of a complex is contained in the pth order betti number, $\beta_p(K)$. The pth order betti number represents the number of p-dimensional holes in a complex and is calculated as the rank of the pth homology group

$$\beta_p(K) = rank\ H_p(K) \ .$$

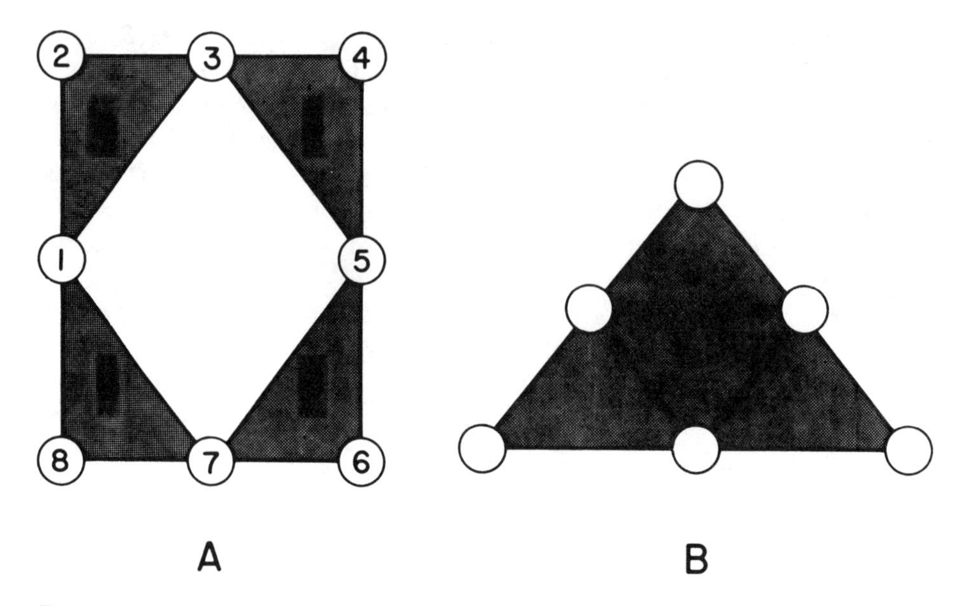

FIGURE 6. Simplicial complexes consisting of 4 species which each use 3 resources. A contains a hole, B is packed more tightly (Sugihara 1982).

This is formed as the quotient group of the family of all p-dimensional cycles, $Z_p(K)$ and the family of bounding cycles $B_p(K)$ in K,

$$H_p(K) = Z_p(K)/B_p(K) \ .$$

$H_p(K)$, therefore, consists of non-bounding p-cycles and the pth order betti number is the number of non-bounding p-dimensional cycles in K. These p-cycles are the kernel of the standard boundary homomorphism.

Some intuition may be gained by demonstrating how to compute the first order betti number for complexes having only 1-dimensional holes (Fig. 7). Put briefly, this involves finding the number of independent 1-cycles [*rank* $Z_1(G)$] in $K(R)$ and subtracting out the number of those that are in the boundary of a species simplex [*rank* $B_1(K)$],

$$\beta_1(K) = Rank\ Z_1(G) - Rank\ B_1(K) \qquad (1)$$

The computation of the first order betti number, therefore, involves finding the number of non-bounding 1-cycles in the 1-skeleton of $K(R)$. Because $G(R)$ is the 1-skeleton of $K(R)$, the number of independent 1-cycles of $K(R)$ is the same as the rank of the cycle basis of $G(R)$.

$$Rank\ Z_1(G) = Rank[cycle\ basis\ G(R)]$$

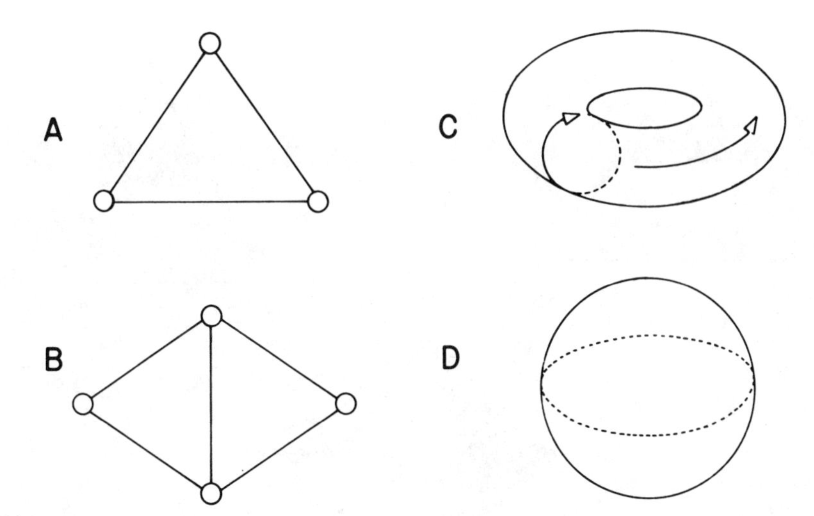

FIGURE 7. Examples of n-dimensional holes. (A) A complex composed of three 1-simplexes framing a single 1-dimensional hole, $\beta_1 = 1$. (B) A complex consisting of five 1-simplexes framing two 1-dimensional holes, $\beta_1 = 1$. (C) A hollow torus containing 2 independent nonbounding 1-cycles, $\beta_1 = 2$. (D) A hollow sphere containing a single 2-dimensional hole, $\beta_2 = 1$ (Sugihara 1982).

The cardinality of the cycle basis for $G(R)$ is calculated from the elegant Euler-Poincare equation

$$Rank\ Z_1(G) = Edges - Vertices + 1 \ . \tag{2}$$

In Figure 8A below, for example, $rank\ Z_1(G) = 9 - 6 + 1 = 4$, hence there are 4 independent 1-cycles in this complex. For 8B and C there are 6 and 8 independent 1-cycles respectively.

The number of independent 1-cycles in the boundaries of all species simplexes [$rank\ B_1(K)$] is calculated from the union of independent 1-cycles in the species cliques in $G(R)$ corresponding to *maximal simplexes* in $K(R)$ (contained no larger simplex). For example, in Figure 8B the maximal simplexes are the 2 shaded triangles and the solid tetrahedron. Applying equation 2 to the cliques corresponding to each maximal simplex (1-skeleton), the number of independent *bounding* 1-cycles is computed as

$$Rank\ B_1(K) = (3 - 3 + 1) + (3 - 3 + 1) + (6 - 4 + 1) = 5 \ .$$

As calculated earlier, the total number of independent 1-cycles (both bounding and non-bounding), $rank\ Z_1(G)$, is 6. Hence applying equation 1 we find

$$\beta_1(K) = 6 - 5 = 1 \ .$$

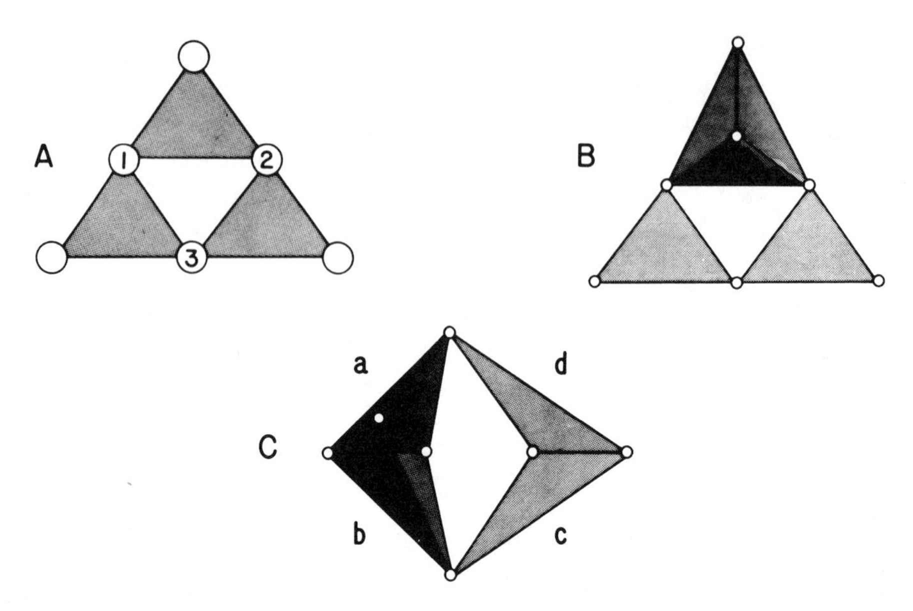

FIGURE 8. Three simplexes containing 1-dimensional holes (non-bounding 1-cycles) (Sugihara 1982).

This indicates the presence of a single 1-dimensional hole in $K(R)$. For simplexes that are joined at 3 or more points and whose intersection thereby contains cycles, rank $B_1(K)$ is computed as the *union* of bounding cycles for each simplex, calculated by working the inclusion-exclusion principle in the usual way to avoid counting cycles more than once. For the complex in Figure 8C, which consists of 2 tetrahedra and 2 triangles, the bounding cycles may be calculated as

$$Rank\ B_1(K) = (3)_a + (3 - 1)_b + (1)_c + (1)_d = 7$$

where 1 is subtracted from b to account for the shared cycle in the intersection of simplexes a and b. In this example, $\beta_1(G) = 8$ so that $\beta_1(K) = 1$, again demonstrating the existence of a single 1-dimensional hole in the complex.

3.1.2 RESULTS

A representative simplicial complex illustrating Bird's (1930) Aspen Parkland community is pictured below (Fig. 9). This community complex is typical in that although it possesses a rich mosaic pattern, from a homological standpoint it is rather simple and contains no holes of dimension 1 or greater. In all,

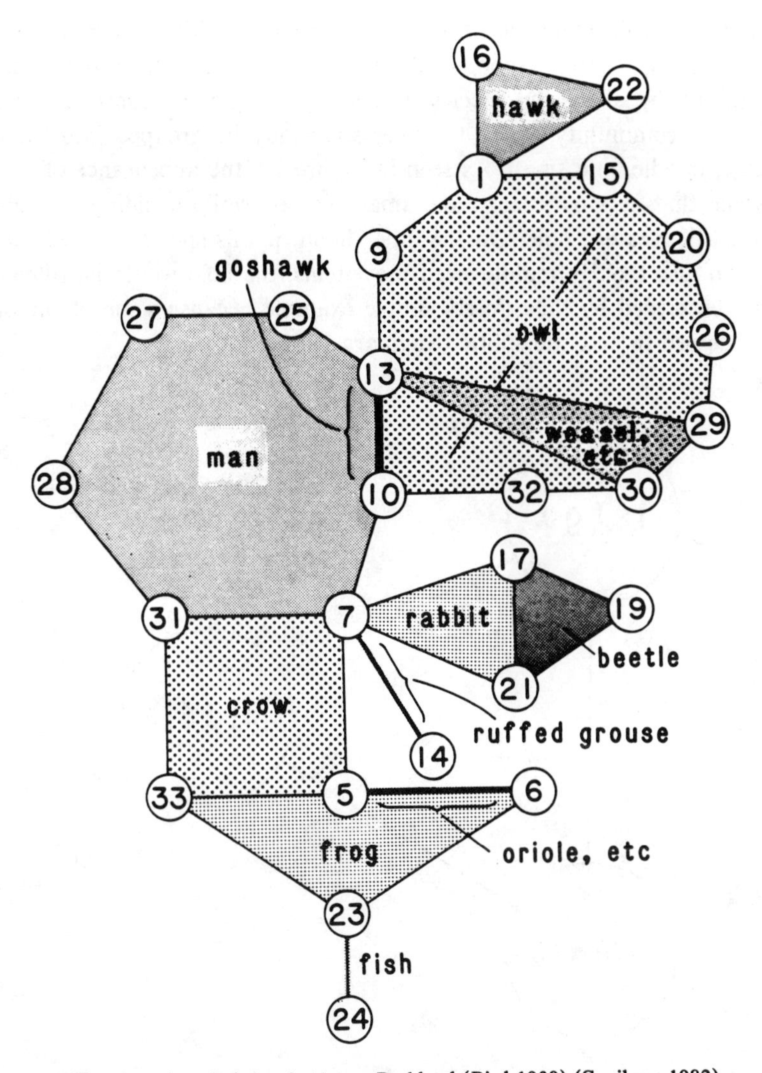

FIGURE 9. $K(R)$ for the Aspen Parkland (Bird 1930) (Sugihara 1982).

out of the the 73 community complexes tested, none had p-dimensional holes with $p \geqslant 2$ bounded by spheres or hyperspheres. Of the 60 communities that could possibly have 1-dimensional holes only two in fact did; although these, it has been argued, should be held with some reserve (Sugihara 1982). The data suggest, therefore, that holes are extremely rare in real niche spaces.

Aside from its interest as a fundamental topological property of communities, the deeper biological significance of this result is the implication that resources in the environment are ordered or correlated with each other by various means (e.g., spatially, taxonomically, by size), and that this ordering is perceived similarly by all of the species involved. For example, consider an insectivorous lizard community where the prey sizes may be grouped into the catagories large, middle, and small. As seen in Figure 10, the appearance of a lizard on the scene that eats only large and small insects, while avoiding the middle-sized ones, will create a hole. On the other hand, if this species obeyed the size ordering and also took in middle-sized prey items, the hole would be filled. One should not be misled by this grossly simple example because some of the orderings are observed to be quite subtle (Sugihara 1982).

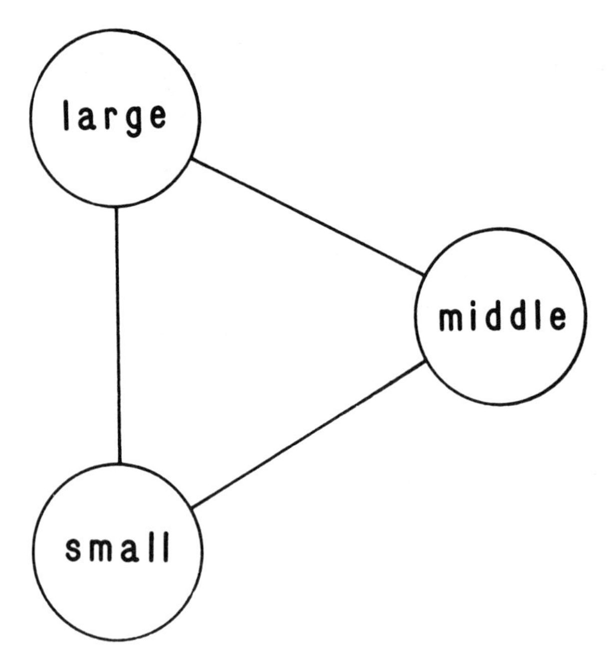

FIGURE 10. $K(R)$ for a community where each species uses 2 resource sizes. Here, the species which uses large and small insects while avoiding the middle-sized ones creates a hole (Sugihara 1982).

3.2 ASSEMBLY RULES FOR G(C)

Turning now to the condensed picture of niche overlaps given by $G(C)$, we will attempt to obtain insight into the process of community assembly from the observation that real consumer overlap graphs possess the rigid circuit property (Sugihara 1982). A connected graph G is said to be ***rigid circuit*** or ***chordal*** if every circuitous path $P_n \epsilon G$ of length $n \geqslant 4$ is shortened by a chord (Dirac

1961) (Fig. 11). Such graphs have also been called *triangulated* (Rose et al. 1967) since all generated subgraphs contain no more than triangular circuits.

The following results for rigid circuit graphs lead to biologically reasonable assembly rules for $G(C)$.

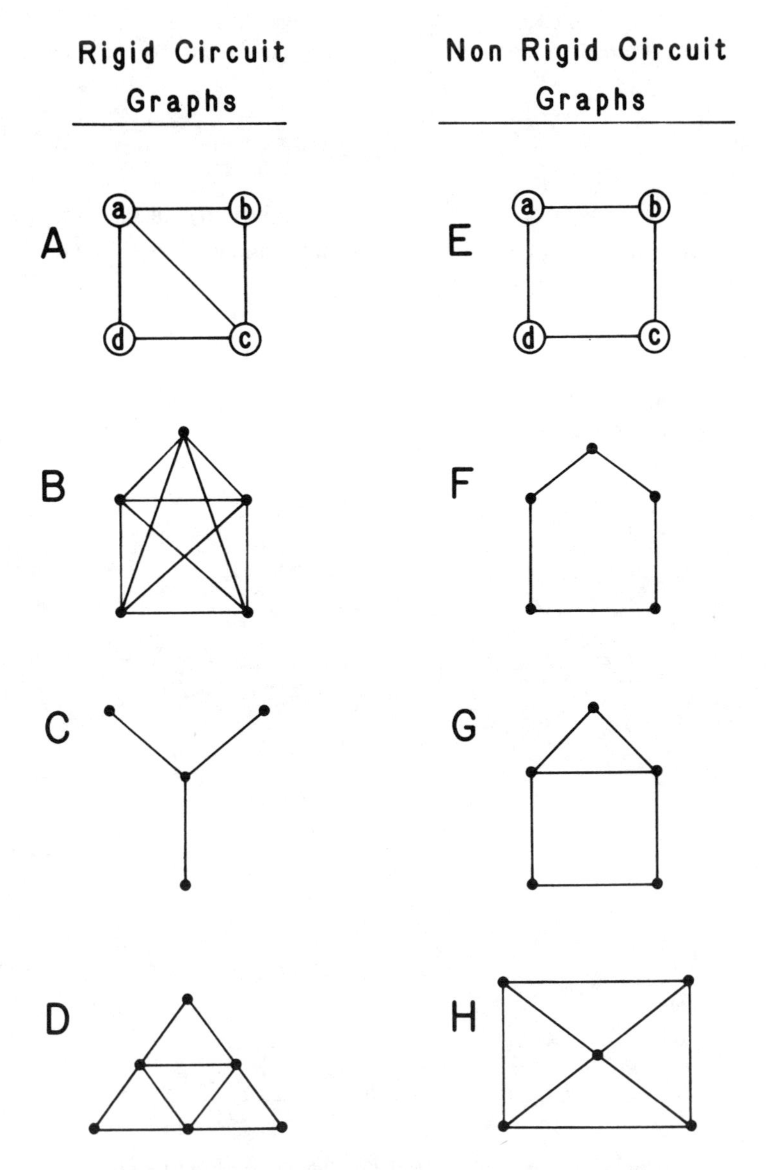

FIGURE 11. An illustration of the rigid circuit property. Notice in graph A, that b and d are extreme vertices (neighbors form a clique) whereas a and c are not (Sugihara 1982).

LEMMA 1. (Dirac 1961). Every rigid circuit graph possesses at least one *extreme vertex*. That is, a vertex whose neighbors form a complete graph or *clique*.

LEMMA 2. If G is a rigid circuit graph, then $G - V$ is also rigid.

PROOF. This follows trivially from the definition of triangulated graphs since every generated subgraph of G must be rigid.

A *perfect elimination ordering* (p.e.o.) (Fig. 12) is an ordering on the vertex set V

$$\alpha: V \rightarrow \{1,2,\ldots,n\}$$

such that for every pair of vertices (v_i,v_j) not joined by an edge there exists no minimal path from v_i to v_j containing only v_i and v_j and vertices numbered less than

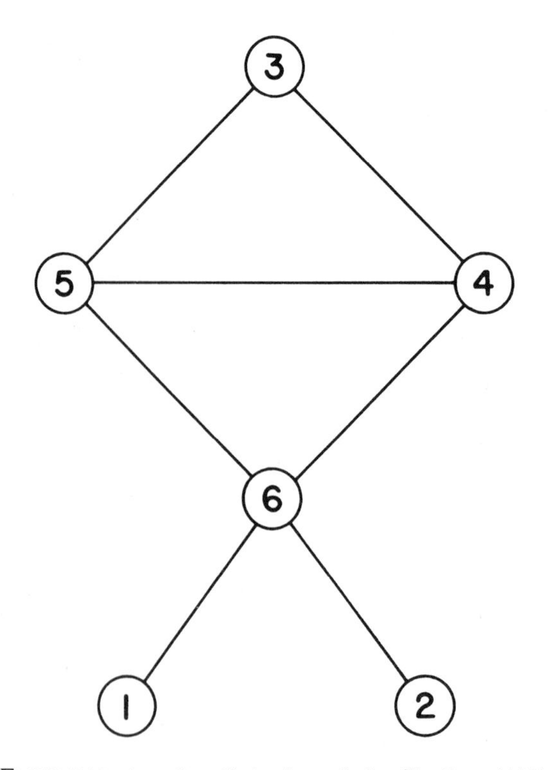

FIGURE 12. A perfect elimination ordering (Sugihara 1982).

$min[\alpha(v_i), \alpha(v_j)]$. Therefore collapsing a graph by following a p.e.o. involves eliminating extreme vertices to generate the sequence of graphs

$$G_n, G_{n-1}, \ldots, G_1$$

having n,...,1 vertices.

THEOREM 1. G is a rigid circuit graph if and only if it possesses a perfect elimination ordering.

PROOF. Lemmas 1 and 2 lead directly to the result that rigid circuit graphs have a p.e.o. Conversely, because a p.e.o. involves eliminating extreme vertices there can be no generated subgraph containing P_n, $n \geqslant 4$ (see also Rose et al. 1976).

The reverse of a perfect elimination ordering is a *perfect addition ordering* (p.a.o.) (Fig. 13),

$$\gamma(V_i) = n - \alpha(V_i) + 1 \ .$$

This generates the sequence of graphs

$$G_1, G_2, \ldots, G_n$$

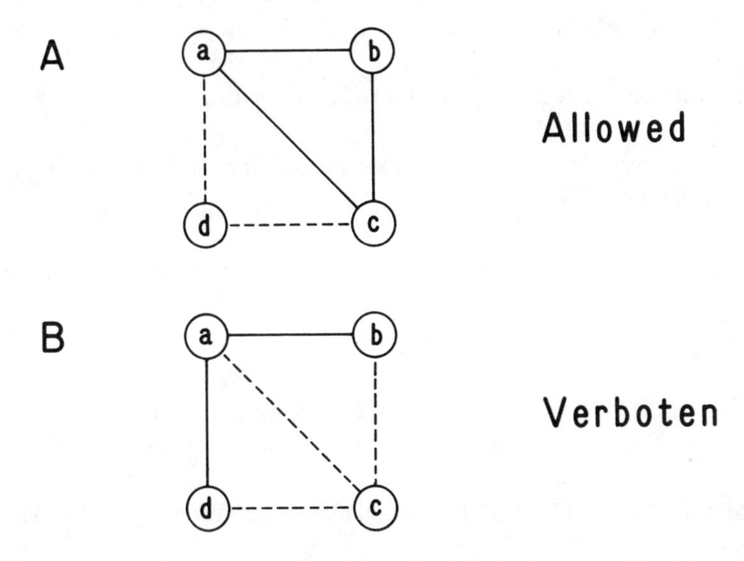

FIGURE 13. An assembly rule deduced from the rigid circuit property (Sugihara 1982).

by adding rather than deleting extreme vertices. Clearly theorem 1 holds for perfect addition as well as perfect elimination.

The fact that real niche overlap graphs *G(C)* are rigid circuit guarantees the existence of a p.a.o. However, to say that such an ordering exists is not to say that real niche overlap graphs were constructed in this way. The following biological arguments may lead to such an ordering.

ARGUMENT 1. It is conventional wisdom that species tend to enter communities in order of increasing (non-decreasing) specialization. That is, generalist species using a wide variety of resources enter first with specialists having narrower requirements coming in later on. Adding sequential specialists to *G(C)* may generate a p.a.o. by requiring incoming species to attach as extreme vertices within single cliques. A clique in *G(C)* corresponds with the biological term *guild*, or group of functionally related species. If an incoming species bridges two different functional groups it will likely be more generalized than any one of the species it overlaps. Therefore, if an incoming species is a specialist it will tend to attach within a single guild, as an extreme vertex. This, in turn generates a p.a.o.

A more rigorous and general argument follows.

ARGUMENT 2. Suppose specialization at the ith stage is associated with fewer niche overlaps or competitors in $G_i(C)$. A specialist, therefore, would have lower point degree (fewer radiating edges) than a generalist. Suppose further that each species subsequent to the ith one is a greater (or equal) specialist with respect to the species already present at the ith stage $(v_1,...,v_{i-1})$. That is, no subsequent species $v_j, j > i$ will overlap more of the vertices $(v_1,...,v_{i-1})$ than v_i. This produces the following rule for ordering vertices in G.

(P1). As the next vertex v_i to number choose the one adjacent to the most labeled vertices $(v_1,...,v_{i-1})$.

To see that this condition generates a p.a.o. we will use the following property of orderings which is strong enough to imply p.e.o. on triangulated graphs (Tarjan 1978).

(P2). If $\alpha(v_o) < \alpha(v_1) < \alpha(v_2), (v_o,v_2) \in E$ and $(v_1,v_2) \notin E$, then there exists a vertex v^* such that $\alpha(v_1) < \alpha(v^*), (v_1,v^*) \in E$ and $(v_o,v^*) \notin E$.

LEMMA 3. (Tarjan 1978). If G is triangulated, any ordering that satisfies *P2* is a p.e.o.

THEOREM 2. If G is triangulated, any ordering generated by *P1* is a p.a.o.

PROOF. Let α be a reverse ordering generated by *P1* where vertices are numbered from n to 1. Due to lemma 3, it will suffice to show that α satisfies *P2*. Suppose $\alpha(v_o) < \alpha(v_1) < \alpha(v_2), (v_o,v_2) \in E$ but $(v_1,v_2) \notin E$. According to *P1* when v_1 is labeled it must be adjacent to at least as many vertices numbered before v_1 as v_o. Since $(v_o,v_2) \in E$ but $(v_1,v_2) \notin E$, there must be some vertex v^* numbered before v_1 to which v_1 is adjacent but not v_o. Thus α satisfies *P2*, and *P1* generates a p.a.o. (see also Tarjan 1978).

Therefore whether one argues loosely as in argument 1 or more precisely as in argument 2, increasing specialization combined with the rigid circuit property leads to the rule that species enter communities by attaching within individual guilds or cliques (perfect addition) rather than across multiple guilds. Given the conditions set forth in argument 2, this rule is both necessary and sufficient.

3.3 ASSEMBLY AND HOLES

The following consequence of the above assembly rule can explain the lack of holes in *K(R)* discussed in Sect. 3.1.

LEMMA 4. Suppose that a guild or a clique in *G(C)* is defined by a common resource v_o in *K(R)*. (True in most observed cases.) Then, perfect addition in *G(C)* pre-empts holes in *K(R)*.

PROOF. It is easy to see that requiring all species; say the ith, to attach to single guilds (p.a.o.) implies that the distance (number of edges) in *K(R)* between v_o and some other resource v_j in the attachment set must be

$$d(v_o,v_j) < 2 \ .$$

This restriction pre-empts the possibility of holes in a community complex.

Therefore rigidity in niche overlaps *G(C)* and the absence of holes may both be explained by our derived assembly rule.

3.4 ASSEMBLY AND INTERVALITY

A graph G is said to be *interval* (Fig. 14) if it is isomorphic to some graph $\Omega(F)$ where F is a family of intervals. That is,

$$v_i \ adj \ v_j \ in \ G \ iff \ f_i \cap f_j \neq \emptyset \ in \ \Omega$$

Intervality is a special case of *boxicity* (Roberts 1969) which is the minimum number of dimensions required of boxes to represent adjacency in *G*. Interval graphs, therefore, have boxicity 1.

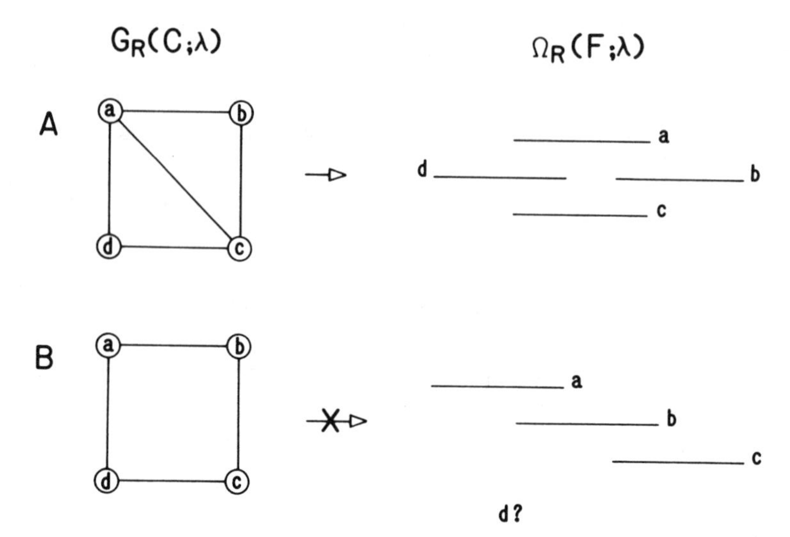

FIGURE 14. Intervality. Graph A can be collapsed to an interval graph, graph B cannot (Sugihara 1982).

The importance of intervality for the study of food webs derives from Cohen's (1978) provocative finding that nature contains an excess of interval niche overlap graphs *G(C)*. To find natural food webs contained in this narrow subset of possible topologies implies a peculiar nonrandomness in the construction of ecological systems. Despite numerous attempts to find an explanation, no biologically plausible mechanism has been discovered for this curious phenomenon (Cohen 1978, Yodzis 1982). One can gain some insight into the problem, however, from the following necessary and sufficient conditions for interval graphs.

THEOREM 3. (Lekerkerker and Boland 1962). A graph G is interval *iff*

1. G is a rigid circuit graph.
2. G is non-asteroidal (i.e., G does not contain 3 distinct points v_o, v_1, v_2 and three paths connecting them W_o, W_1, W_2 such that each point v_i of this triple of points is not a neighbor of any path W_i connecting the other two).

Because real niche overlap graphs are rigid circuit it seems worthwhile to test whether conditioning on rigidity alone may account for the high frequency of interval niche overlap graphs. That is, could the excess of interval niche overlap graphs in nature be a simple consequence of the fact that they are triangulated? In particular could it be a consequence of the assembly rule for rigid circuit graphs deduced earlier?

Out of our collection of 73 nontrivial communities *G(C)* (connected components with 2 or more species), 63 were observed to be interval. When these communities are numerically assembled by a random perfect addition ordering, 62.36 are found to be interval, and the difference is not significant ($z = 0.33$). These random rigid circuit graphs have the same number of vertices and edges as the observed communities however they are assembled by a perfect addition ordering which uses uniform distributions over appropriate limits to determine the sizes and identities of the attachment sets at each step. Therefore the assembly rule deduced in Sect. 3.2 appears to be sufficient to account for the high frequency of interval niche overlap graphs observed in nature.

4. REFERENCES

[1] Atkin, R. H., *Mathematical Structure in Human Affairs.* Crane-Russak, N.Y., N.Y. (1974).

[2] Bird, R. D., Biotic communities of the aspen parkland of Central Canada. Ecology 11 (1930), 356–442.

[3] Boesch, F., Introduction to basic network problems. Lecture Notes: The Mathematics of Networks, AMS Short Course Series, 1981.

[4] Briand, F., Environmental control of food web structure, Ecology (1983).

[5] Cohen, J. E., *Food Webs and Niche Space.* Princeton Univ., Press, Princeton, 1978.

[6] Day, J. H., The biology of the Knysna Estuary, South Africa. In *Estuaries.* G. H. Lauff, ed. (AAAS publication 83) Washington (1967), 397–404.

[7] Dirac, G. A., On rigid circuit graphs, Abh, Math, Sem. Univ. Hamburg 25 (1961), 71–76.

[8] Giblin, P. J., *Graphs Surfaces and Homology,* Chapman and Hall, London (1977).

[9] Harary, F., *Graph Theory,* Addison Wesley (1972).

[10] Roberts, F. S., *Discrete Mathematical Models, with Applications to Social, Biological and Environmental Problems.* Prentice Hall, N.J. (1976).

[11] Rose, D. J., R. E. Tarjan, G. S. Leuker, Algorithmic aspects of vertex elimination on graphs, SIAM J. Comput. 5 (1976), 266–283.

[12] Sugihara, G., *Niche Hierarchy: Structure, Organization, and Assembly in Natural Communities.* Ph.D. Dissertation, Princeton Univ. (1982).

[13] Tarjan, R. E., Private manuscript, Bell Labs Technical Course.

ENVIRONMENTAL SCIENCES DIVISION
OAK RIDGE NATIONAL LABORATORY
OAK RIDGE, TENNESSEE 37830

ABCDEFGHIJ–CM–8987654